by Gary Imm

Introduction

This book is my view of the best 50 objects to be seen in the night sky from the northern hemisphere.

What are the most magnificent wonders of the universe? Many lists of wonders have been created:
- The **7 classic wonders** include the Pyramids, the Colossus of Rhodes, and the Hanging Gardens.
- The **7 engineering wonders** include the Empire State Building, the Golden Gate Bridge, and the Panama Canal.
- The **7 natural wonders** include the Grand Canyon, Mount Everest, and the Great Barrier Reef.
- The **7 new wonders** include the Great Wall of China, the Colosseum, and Machu Picchu.

But for me, nothing listed above compares to the beauty, scope, mystery and awe of the wonders of the heavens shown in this book. These wonders have been hidden from our eyes for the vast majority of human history. It has only been in the past few years that we can see these sights from our own backyards.

All of the images shown here were obtained from my temporary backyard setup in Onalaska, Texas. The images were processed to enhance detail and color but no content was artificially created.

There are 9 different types of space objects captured here - dark nebulae, emission nebulae, planetary nebulae, reflection nebulae, molecular clouds, supernova remnants, open star clusters, globular star clusters, and galaxies.

While the stars and galaxies of this book are captured in "true color" images, the gaseous nebulae are much fainter, so narrowband gas filters have been used to capture the faint detail. Typically, a SHO color palette is used, where Sulfur is assigned to the red channel, Hydrogen to the green channel, and Oxygen to the blue channel. This is referred to as the Hubble palette, made famous by the scientists operating the Hubble telescope.

The 50 objects are listed in order of increasing distance from us, measured in light years. The objects that are further away from us are typically larger in actual size as well. Light years are used as the unit of measure for both distance and size, since space is so vast in scale that using miles becomes unwieldy. For example, the closest object to us in this book is the Rho Ophiuchi Nebula, which is 360 light years away. That is equivalent to 2,116,305,134,345,916 miles away. Using light years keeps the number size more manageable.

Each full color high resolution object image is accompanied by a description, including the location, distance and size of the object. Acquisition details are also provided, including the equipment used to capture the image and the exposure time. The objects are listed below by page number, using both their nickname and their official designation.

Images

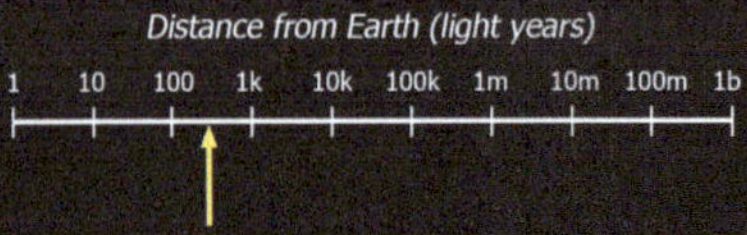

Object Types Included

X Nebula (Dark)
X Nebula (Emission)
   Nebula (Molecular Cloud)
   Nebula (PN)
X Nebula (Reflection)
   Nebula (SNR)
   Star Cluster (Open)
X Star Cluster (Globular)
   Galaxy

Rho Ophiuchi Nebula

Object:

Location Constellations of Scorpius and Cassiopeia, at a declination of -25 degrees

Distance 360 light years (distance of IC 4603, the closest of all of the objects in the image)

Size 20 light years

Description This region is one of the most naturally colorful in the sky and the closest major nebula to earth.

Unlike the many narrowband nebulae images in this book which use gas filters, this galaxy image is taken with simple luminance (clear), red, green, and blue filters. The image consists of many different objects at different distances, all superimposed in the same space.

At the bottom left of the image is the intense orange bright star Antares. At magnitude 1, this red supergiant is the 15th brightest star in the sky. Its name is from the Greek, meaning "rival to Mars". It is located about 600 light years from us.

To the right of Antares is M4, the closest globular cluster to us in the entire sky at a distance of 6000 light years. This is the first cluster in which individual stars were resolved. M4 is about 30 arc-minutes wide and 60 light years in diameter. It consists of about 100,000 stars.

I find it interesting that these 2 objects at the bottom of the page - the orange star Antares on the left and the white star cluster M4 on the right - look to be about the same size in the image, but Antares is a only single star and M4 has 100,000 stars. But M4 is much further away.

Just above and right of Antares is the globular star cluster NGC 6144. This cluster is much further away than M4, at 33,000 light years. Although it looks much smaller than M4, it is about the same actual size.

Above Antares, at the center left of the image, is the blue reflection nebula, IC 4605. It surrounds the star 22 Scorpii.

The bright blue star on the right side of the image is Sigma Scorpii. This magnitude 2.9 star, located 700 light years away, is ionizing the SH2-9 red hydrogen emission gas front. This star is known as Alniyat, which is Arabic for "the arteries". This refers to its proximity to Antares, the Scorpion's Heart.

Just above image center is one of the most interesting reflection nebulae in the sky, IC 4603, located 360 light years away. It is located closer to us than anything else here. In different images, this nebula can have aspects of white, blue, yellow, and purple color. The dark nebulae patterns around this object are interesting.

3 Barnard dark nebulae are seen in the image. Barnard 44 extends left from IC 4605, while Barnard 45 extends up and left from IC 4603. Surrounding IC 4603 is Barnard 42.

At the top of the image is the triple star system, Rho Ophiuchi, the only part of this image that lies in the constellation of Ophiuchus. Rho Ophiuchi is surrounded by a blue reflection nebula, designated as IC 4604.

Image Acquisition:

Date June 25, 2023

Exp. Time 1 hour

Telescope Celestron RASA 11" V2
Camera ZWO ASI6200MM Pro
Mount Astro-Physics Mach1 GTO
Filters Broadband - Lum, Red, Green & Blue

"My own hand laid the foundation of the earth, and my right hand spread out the heavens." - Isaiah 48:13

Object Types Included

   Nebula (Dark)
   Nebula (Emission)
X Nebula (Molecular Cloud)
   Nebula (PN)
   Nebula (Reflection)
   Nebula (SNR)
   Star Cluster (Open)
   Star Cluster (Globular)
   Galaxy

Baby Eagle Nebula (LBN 777)

Object:

Location   Constellation of Taurus, at a declination of +26 degrees

Distance   400 light years

Size   3 light years

Description   This object is a very faint molecular cloud known as the Baby Eagle Nebula and the Vulture Head Nebula, plus various other bird head creative nicknames. Its formal designation is LBN 777. It is close to the Pleiades star cluster in a large cloud of dust and gas known as the Taurus Molecular Cloud.

Note the tremendous amount of dust seen throughout the image. This dust is the same density as the dust in the room that you sleep in, but these dust clouds are light years thick, which is enough to block out the light of the stars behind it. You will see many more dust clouds in this book - space is filled with more dust than most people realize.

The molecular cloud envelopes a Bok globule, known as Bernard 207. This Bok globule is the brownish patch to the right of the Eagle's eye (the "brain"). The brownish color is caused by large dust grains embedded in the gas. Bok globules are isolated small dark nebulae containing dense dust and gas, within which star formation is believed to be taking place. Studies have concluded that the age of the globule is from 100,000 to 300,000 years.

Just above the cloud is the bright orange magnitude 8.3 star, HD 25596. This star is much further away than the cloud, at a distance of 2200 light years. Early in this book, this is a good place to see that 2 objects which *look* close together - the orange star and the cloud - are actually very far apart. Unlike on earth which the landscape gives us depth perspective, space offers our eyes nothing to help us see a distance perspective when looking at adjacent objects. Many times in this book you will be tempted to make the assumption that 2 adjacent objects, like 2 galaxies, or a star and a nebula, are physically close to each other, when most often they are not.

My favorite part of the image is the faint thin arc of bright gas which curves around the top of the cloud and through the bright orange star. I haven't seen an explanation for this interesting feature.   Despite its appearance, the star is too far away to have any effect on the nebula.

Image Acquisition:

Date   Nov 28, 2022

Exp. Time   2 hours

Telescope   Celestron RASA 11" V2
Camera   ZWO ASI6200MM Pro
Mount   Astro-Physics Mach1 GTO
Filters   Broadband - Lum, Red, Green & Blue

"There are more stars in the heavens than all the grains of sands covering the world's beaches"
– Carl Sagan

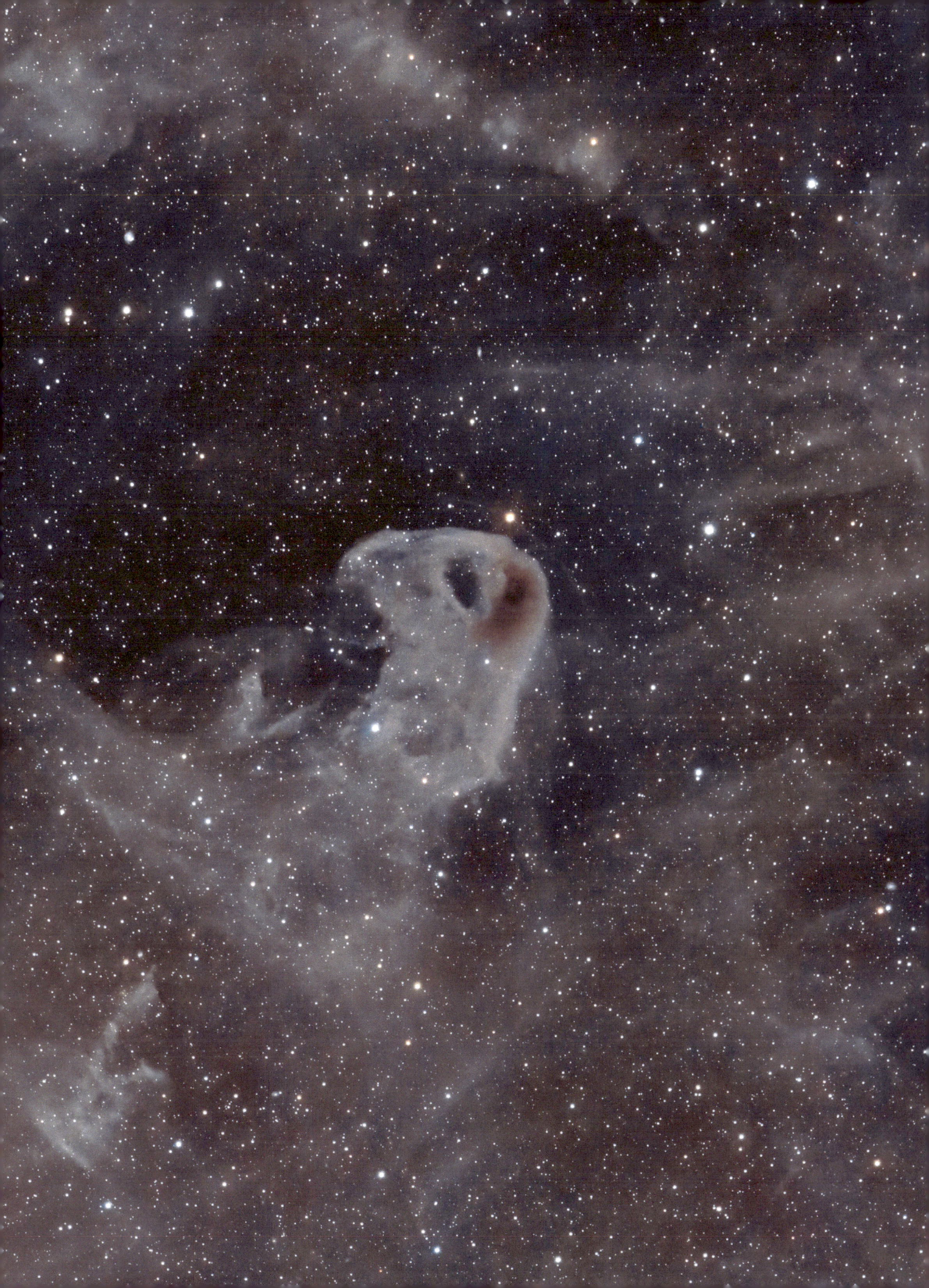

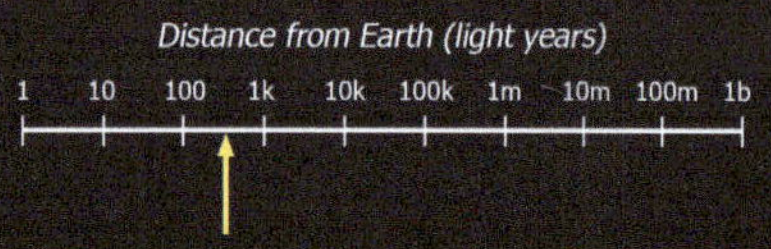

Object Types Included

Nebula (Dark)
Nebula (Emission)
Nebula (Molecular Cloud)
Nebula (PN)
X Nebula (Reflection)
Nebula (SNR)
X Star Cluster (Open)
Star Cluster (Globular)
Galaxy

Pleiades (M45)

Object:

Location Constellation of Taurus, at a declination of +24 degrees

Distance 400 light years

Size 10 light years

Description This popular open star cluster, nicknamed the Pleiades and the Seven Sisters, is one of the brightest and closest open clusters in the sky, containing over 3000 stars. It spans 1.2 degrees in our apparent view, which corresponds to an actual width of about 10 light years.

This is the most famous star cluster in the sky, easily visible with the naked eye and the brightest object in the entire Messier Catalog. Many people, including myself for many years, mistakenly refer to this cluster as the Little Dipper. The real Little Dipper originates at the North Star, is barely visible, and is much bigger in size. This cluster looks sort of like a tiny dipper with a very short handle.

The nickname of this cluster is the Seven Sisters, even though only six bright stars are easily visible to the naked eye (in the darkest skies, some people can see as many as 14!). The Japanese name for this cluster is Subaru, which explains why the Pleiades star pattern is the basis for the Subaru car logo. The Bible makes 3 references to Pleiades – in Amos 5:8, Job 9:7-9, and Job 38:31-33.

The Pleiades has always been a favorite object of mine, dating back to my early visual observing times. Even now, I see how many Pleiades stars I can see with my naked eye, even though it seems I am seeing less as I get older.

Although the stars here are the main focus, what really makes this object special is the magnificent dust cloud around the stars which is reflecting the blue star light and creating beautiful reflection nebulae. I am fascinated by all of the different nebulosity shapes - straight lines, long curving arcs, billowing clouds, shimmering sheets, and areas of fantastic chaos.

I initially assumed that the cloud around the stars is leftover dust from the original star cluster formation, which is sometimes seen in other clusters. But based on velocity measurements, scientists believe that the reflection nebulae are due to a molecular cloud that is unrelated to the star cluster. The cluster and cloud are simply crossing paths.   A long time from now, all that will be seen here are bright stars.

Image Acquisition:

Date Dec 23, 2022

Exp. Time 2 hours

Telescope Celestron RASA 11" V2
Camera ZWO ASI6200MM Pro
Mount Astro-Physics Mach1 GTO
Filters Broadband - Lum, Red, Green & Blue

"Can you bind the chains of the Pleiades?" - Job 38:31

Object Types Included

X Nebula (Dark)
   Nebula (Emission)
   Nebula (Molecular Cloud)
   Nebula (PN)
X Nebula (Reflection)
   Nebula (SNR)
   Star Cluster (Open)
X Star Cluster (Globular)
   Galaxy

Anteater Nebula (NGC 6726)

Object:

Location   Constellation of Corona Australis (The Southern Crown), at a declination of -37 degrees

Distance   400 light years

Size   20 light years

Description   The dust cloud is shaped like an anteater with a huge snout lumbering to the right across the sky. To see it in this image, you need to turn your head to the left and use your imagination.

The dark nebula is Bernes 157, is one of the dustiest, darkest nebulae in the sky. The dust clouds block many of the more distant background stars in the Milky Way.

Straddling the dark nebula are two wonderful, detailed blue reflection nebulae. The left blue nebula actually consists of two objects, NGC 6726 and 6727, separated by 1 arc-minute. The nebulae are powered by 2 young stars shrouded in gas and dust, likely having protoplanetary disks. The right blue nebula is the equally interesting object IC 4812, illuminated by a double star of magnitudes 6.4 and 6.7. These two stars have a separation of only 10 arc-seconds. The closeup below shows these blue reflection nebulae in more detail.

A number of smaller white reflection nebulae are seen in the vicinity. The largest of these is NGC 6729, seen at bottom center of the closeup below. Like Hubble's Variable Nebula, this object consists of dark clouds obscuring the light of a bright star, in this case the star R Coronae Australis. This variable star can change in brightness almost 4 magnitudes due to the moving cloud shadows, with observable changes in as short of a time as 24 hours.

Finally, the spectacular globular cluster NGC 6723 is seen at the upper left of the main image on the opposite page. This cluster has a magnitude of 6.4 and a diameter of 10 arc-minutes. It is much further away than the nebula complex, at a distance of 30,000 light years. The cluster is believe to be comprised of about 0.5 million stars.

Image Acquisition:

Date   July 11, 2023

Exp. Time   2 hours

Telescope   Celestron RASA 11" V2
Camera   ZWO ASI6200MM Pro
Mount   Astro-Physics Mach1 GTO
Filters   Broadband - Lum, Red, Green & Blue

Closeup of Blue Reflection Nebulae:

"Lift your eyes and look to the heavens: who created all these? He who brings out the starry host one by one, and calls them each by name. Because of his great power and mighty strength, not one of them is missing."
- Isaiah 40:26

Object Types Included

Nebula (Dark)
Nebula (Emission)
X Nebula (Molecular Cloud)
Nebula (PN)
X Nebula (Reflection)
Nebula (SNR)
Star Cluster (Open)
Star Cluster (Globular)
Galaxy

Boogeyman Nebula (LDN 1622)

Object:

Location   Constellation of Orion, at a declination of +2 degrees

Distance   500 light years

Size   10 light years

Description   This object is a cometary molecular cloud. The densest parts of this cloud completely block out the background stars.

The most interesting aspect of this object to me is the bright yellow young pre-main sequence star (HBC 515) seen on the lower right side of the image in the head of the cloud. I have included a closeup image below. This star is one of the visually brightest young stars known. The star is surrounded by a yellow reflection nebula designated as vdB 62 or Parsamian 3. I also find it interesting that the cloud is silhouetted against a faint background of glowing reddish hydrogen gas.

This cloud has been identified as the birthplace of many other stars besides HBC 515. Typical of such a cloud, a number of Herbig-Haro objects and outflows have been identified in this region but I could not positively identify many of these very faint objects in my image.

I am not a big fan of nicknames. But this object's nickname, the Boogeyman Nebula, seems fitting for its dark, foreboding look.

Image Acquisition:

Date   Dec 25, 2022

Exp. Time   2 hours

Telescope   Celestron RASA 11" V2
Camera   ZWO ASI6200MM Pro
Mount   Astro-Physics Mach1 GTO
Filters   Broadband - Lum, Red, Green & Blue

Closeup of Boogeyman's Head:

"The strongest affection and utmost zeal should, I think, promote the studies concerned with the most beautiful objects. This is the discipline that deals with the universe's divine revolutions, the stars' motions, sizes, distances, risings and settings . . . for what is more beautiful than heaven?" - - Nicolaus Copernicus, Astronomer

Object Types Included
Nebula (Dark)
Nebula (Emission)
Nebula (Molecular Cloud)
X   Nebula (PN)
Nebula (Reflection)
Nebula (SNR)
Star Cluster (Open)
Star Cluster (Globular)
Galaxy

Eye of God Nebula (NGC 7293)

Object:

Location   Constellation of Aquarius, at a declination of -21 degrees

Distance   650 light years

Size   3 light years

Description   This object, also known as the Helix Nebula, is a planetary nebula. It is one of the closest to Earth of all the bright planetary nebulae. Its apparent size is about half the diameter of the full moon. This nebula is the largest of the 250 planetary nebula that I have imaged.

Planetary nebula are intermediate sized stars, like our sun, which shed and illuminate their outer layers of gas near the end of their life. The red in this image corresponds to hydrogen and the blue to oxygen.

The source star in this case is the small central bluish star, which over time is destined to evolve into a tiny white dwarf. This star is visible in the image as the pinpoint of light at the exact center of the blue region.

This object has many radial filaments around the circumference, all pointing back to the central star. These filaments are likely gaseous streams ejected from this star. Many of these filaments resolve into cometary knots. Some of these knots are visible in the image as tiny puffs of white clouds along the outer edge of the bluish core. The knots look tiny but each are larger than our solar system.

Many different explanations have been proposed over the years for the morphological structure of this nebula, including 2 overlapping rings or a doughnut. The one which makes the most sense to me is a bi-polar structure with 2 spherical lobes (like an hourglass) expanding from the central star, constricted by a toroidal ring of dust and gas around its equator. The white region, appearing as almost a circular ring in the image, is the part of the nebula constricted by the toroidal ring.

The major axis of the nebula is oriented from 5 o'clock to 11 o'clock. We are looking at the nebula at an angle of about 35 degrees from end-on. There is some lobe breakout of gas at each end of the nebula, where the reddish outer nebula is seen to come to a point. This helps gives this nebula its distinctive "eye" shape.

Image Acquisition:

Date   Aug 26, 2020

Exp. Time   14 hours

Telescope   Celestron EdgeHD 11"
Camera   ZWO ASI6200MM Pro
Mount   Astro-Physics Mach1 GTO
Filters   Broadband (for stars) - Red, Green & Blue
Narrowband (for nebula) - HII & OIII
The narrowband data was mapped as follows - HII to red, OIII to green & blue (HOO palette).

"When life gets too overwhelming, just look up at the night sky and lose yourself for a while."
- Deborah a Ten Brink

Object Types Included

X   Nebula (Dark)
    Nebula (Emission)
X   Nebula (Molecular Cloud)
    Nebula (PN)
X   Nebula (Reflection)
    Nebula (SNR)
    Star Cluster (Open)
    Star Cluster (Globular)
    Galaxy

Shark Nebula (LBN 535)

Object:

Location Constellation of Cepheus, at a declination of +73 degrees

Distance 1000 light years

Size 40 light years

Description This image captures a dark nebulae, a molecular cloud, two reflection nebulae, and a galaxy. The nebulae are fairly close to us inside of the Milky Way at about 1000 light years distant, while the galaxy is about 70,000 times further away.

Collectively these objects are known as the Shark Nebula, seen diving down towards the left corner of the image.

The body of the shark is the largest object, the molecular cloud LBN 535. The dark nebula LDN 1235 makes up the shark's snout. Directly above the eye, vdB 150 is the bright bluish reflection nebula on top of the shark's head. At the shark's chest is another bright bluish reflection nebula, vdB 149.

Towards the top of the image, the white spiral galaxy UGC 11861 is located above the bright white star 16 Cep. This is a magnitude 14 galaxy located 70 million light years away which spans 3 arc-minutes in our apparent view. This corresponds to a diameter of 70,000 light years, just over half the size of our own Milky Way galaxy.

Image Acquisition:

Date July 28, 2023

Exp. Time 2 hours

Telescope Celestron RASA 11" V2
Camera ZWO ASI6200MM Pro
Mount Astro-Physics Mach1 GTO
Filters Broadband - Lum, Red, Green & Blue

"Shoot for the stars, but if you happen to miss, shoot for the moon instead."
- Neil Armstrong

Object Types Included

    Nebula (Dark)
X   Nebula (Emission)
X   Nebula (Molecular Cloud)
    Nebula (PN)
X   Nebula (Reflection)
    Nebula (SNR)
    Star Cluster (Open)
    Star Cluster (Globular)
    Galaxy

Embryo Nebula (NGC 1333)

Object:

Location   Constellation of Perseus, at a declination of +31 degrees

Distance   1100 light years

Size   4 light years

Description   This colorful, complex object is one of the nearest and most active star formation regions in the sky. The field of view includes a combination of reflection nebulae, emission nebulae, and emerging stars, all within a molecular cloud that spans 4 light years. The molecular cloud is relatively young, containing hundreds of forming stars. Most of the stars are hidden to us by the dust and are only identified through infrared.

I love the subtly different shadings of reds, yellows, oranges, blues and whites of the nebula and surrounding stars, together with the contrast of the dark nebula clouds. The primary blue object is vdB 17, a reflection nebula surrounding the bright star BD +30 549. The yellow objects are reflection nebulae surrounding older stars, while the red emission nebulae are outflows from protostars forming within the cloud.

So why is this object called the Embryo Nebula? I have not found a definitive explanation. It doesn't look like any kind of embryo to me. I think the nickname is due to this cloud complex being a massive incubator for new stars.

Image Acquisition:

Date   Dec 14, 2022

Exp. Time   2 hours

Telescope   Celestron RASA 11" V2
Camera   ZWO ASI6200MM Pro
Mount   Astro-Physics Mach1 GTO
Filters   Broadband - Lum, Red, Green & Blue

"If people sat outside and looked at the stars each night, I'll bet they'd live a lot differently."
- Bill Watterson (creator of Calvin & Hobbes)

Object Types Included

X Nebula (Dark)
  Nebula (Emission)
  Nebula (Molecular Cloud)
  Nebula (PN)
X Nebula (Reflection)
  Nebula (SNR)
  Star Cluster (Open)
  Star Cluster (Globular)
  Galaxy

Wolf's Cave Nebula (vdB 152)

Object:

Location   Constellation of Cepheus, at a declination of +71 degrees

Distance   1200 light years

Size   4 light years

Description   This object is a 2 degree long combination of a blue reflection nebula (vdB 152) and a much longer dark nebulae (Barnard 175 or LDN 1217).

This object has 2 nicknames – the Cepheus Flare and Wolf's Cave. Although the latter nickname suggests that this sky object resembles a wolf's cave, this object is actually named after Max Wolf, the astronomer who announced its discovery in 1908.

vdB 152, also known as Cederblad 201, spans about 4 light years, while Barnard 175 spans about 60 light-years.

The young star illuminating vdB 152 is the 9.3 magnitude blue star BD+69 1231. Surprisingly, this star did not form in this cloud – it is just passing through. Its measured velocity is much different than the cloud's velocity. My favorite part of this image are the interesting arcs of blue light which surround this blue star.

The colorful nebula at the right edge of the image is DeHt5. This object is a very faint, very old planetary nebula (PN) located just over 1000 light years away. The bluish star in the center of the purple-blue region is the central star.

The faint red ribbon in the upper right corner of the image arcing through vdB 152 is the supernova remnant (SNR) G110.3 + 11.3. It is one of the nearest SNRs to earth, at only 1300 light years away.

Image Acquisition:

Date   July 24, 2023

Exp. Time   2 hours

Telescope   Celestron RASA 11" V2
Camera   ZWO ASI6200MM Pro
Mount   Astro-Physics Mach1 GTO
Filters   Broadband - Lum, Red, Green & Blue

Closeup of vdB 152:

"Astronomy compels the soul to look upward, and leads us from this world to another."
- Plato, Philosopher

Object Types Included

Nebula (Dark)
Nebula (Emission)
Nebula (Molecular Cloud)
Nebula (PN)
X Nebula (Reflection)
Nebula (SNR)
Star Cluster (Open)
Star Cluster (Globular)
Galaxy

Iris Nebula (NGC 7023)

Object:

Location   Constellation of Cepheus, at a declination of +68 degrees

Distance   1400 light years

Size   6 light years

Description   This object, also known as vdB 139, is a beautiful blue reflection nebula. A bright 7th magnitude star (HD 200775) illuminates the core of the nebula. The pretty blue "petals" of the Iris Nebula span about six light-years. A closeup of the nebula is shown below.

The dominant color of the reflection nebula is blue, which is true for most reflection nebulae. The typical size of dust grains in a cloud is comparable to the wavelength of blue light. As a result, blue light is scattered more efficiently than longer, red wavelengths giving the characteristic blue color for typical reflection nebulae.

However, often (as in this case) many other color shades are also apparent. The subtle color variations in this nebula include purple, brown, black, and white.

I find interesting that "windows" seem to be carved through different parts of the cloud, especially towards the right of center where the background sky can be fully seen. I don't understand why there are such distinct boundaries. Perhaps these windows are partially due to local stellar jets clearing out areas of dust.

I like the bright white "arcs" at the center of the nebula, generally running almost horizontally. I imagine that these are cloud fronts illuminated by bright star light.

The small white reflection nebula seen just left and slightly below the nebula is the young stellar object [KWD2009] 57.

Image Acquisition:

Date   Oct 14, 2023

Exp. Time   2 hours

Telescope   Celestron RASA 11" V2
Camera   ZWO ASI6200MM Pro
Mount   Astro-Physics Mach1 GTO
Filters   Broadband - Lum, Red, Green & Blue

Closeup of Central Region:

"A star does not compete with other stars around it; it just shines."
— Matshona Dhliwayo

Nebula (Dark)
Nebula (Emission)
Nebula (Molecular Cloud)
Nebula (PN)
Nebula (Reflection)
X Nebula (SNR)
Star Cluster (Open)
Star Cluster (Globular)
Galaxy

Cygnus Loop

Object:

Location   Constellation of Cygnus, at a declination of +31 degrees

Distance   1500 light years

Size   80 light years

Description   This object is the famous Cygnus Loop, a large supernova remnant (SNR) also known as the Veil Nebula Complex. This SNR includes the Eastern Veil and Southeastern Knot on the left, Pickering's Triangle and NGC 6974 at the top, and the Witch's Broom Nebula (Western Veil) on the right.

The bright foreground star near the center of the right edge of the image is 52 Cygnus. This magnitude 4 star is only 200 light years away and is not associated with the supernova remnant.

Most stars die a quiet death, as in a planetary nebula, but a select few massive stars end their life in a more dramatic way, through a powerful supernova explosion.

Astronomers estimate that the supernova explosion that produced this nebula occurred between 3,000 BC and 6,000 BC. The image covers an area of about 36 times the size of the full moon.

The spherical nature of the expanding supernova remnant is easy to see in this image. The source star of this supernova explosion has not been found. It may have been destroyed in the explosion.

The spindly filaments of hydrogen and oxygen that outline the gas shock fronts throughout this region are one of my favorite DSO objects.

Image Acquisition:

Date   July 24-25, 2023

Exp. Time   5 hours

Telescope   Celestron RASA 11" V2
Camera   ZWO ASI6200MM Pro
Mount   Astro-Physics Mach1 GTO
Filters   Broadband (for stars) - Red, Green & Blue
Narrowband (for nebula) - HII & OIII
The narrowband data was mapped as follows - HII to red, OIII to green & blue (HOO palette).

Closeup of Witch's Broom Nebula (on right side, rotated 90 degrees):

"You alone are the Lord. You made the heavens, even the highest heavens, and all their starry host."

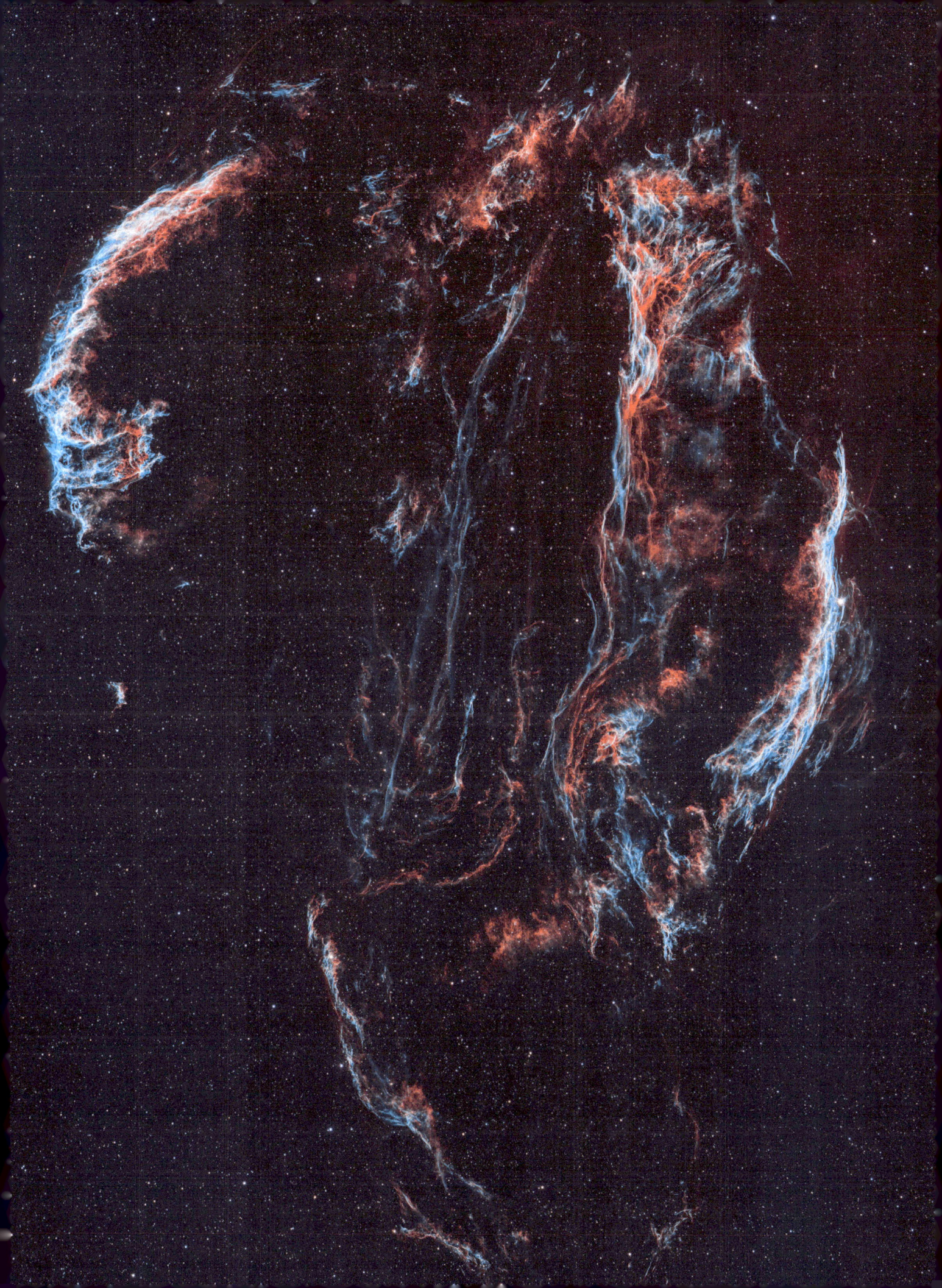

Object Types Included

Nebula (Dark)
X Nebula (Emission)
Nebula (Molecular Cloud)
Nebula (PN)
Nebula (Reflection)
Nebula (SNR)
Star Cluster (Open)
Star Cluster (Globular)
Galaxy

North American Nebula (NGC 7000)

Object:

Location	Constellation of Cygnus, at a declination of +44 degrees
Distance	1500 light years
Size	90 light years

Description The North America nebula is one of the largest emission nebulae in the sky. The Gulf of Mexico shape is defined in black, the only place in the image where the bright nebula is not present.

This object is the first of many emission nebulae in this book. They are the most colorful objects in the sky. An emission nebula is a type of astronomical object made up of ionized gases that emit light of various colors. Essentially, it's a big cloud of gas and dust in space that glows, due to ionized gases that emit light of various colors. The glow occurs when the nebula's atoms are excited by high energy ultraviolet light from nearby young, hot stars.

This image is a complex scene with objects that are at various distances from us. The large bright foreground Milky Way stars are closest. Next at about 1500 light years is the dark molecular cloud which partially lies in front of the emission nebula. Next is the emission nebula itself at about 2000 light years. Finally, many dimmer stars are seen in the more distant background.

The most striking feature of this nebula is the prominent emission ridge towards the bottom left of the image known as the Cygnus Wall. A closeup of the wall is shown below. The wall has been sculpted by radiation from the region's young, hot, massive stars and is an area of active star formation.

Image Acquisition:

Date	Oct 17, 2023
Exp. Time	3.5 hours
Telescope	Celestron RASA 11" V2
Camera	ZWO ASI6200MM Pro
Mount	Astro-Physics Mach1 GTO
Filters	Broadband (for stars) - Red, Green & Blue
	Narrowband (for nebula) - HII, OIII & SII
	The narrowband data was mapped as follows - SII to red, HII to green & OIII to blue (SHO palette).

Closeup of the Cygnus Wall:

"When you reach for the stars you may not quite get one, but you won't come up with a handful of mud either."
– Leo Burnett

Object Types Included

X Nebula (Dark)
   Nebula (Emission)
   Nebula (Molecular Cloud)
   Nebula (PN)
   Nebula (Reflection)
   Nebula (SNR)
   Star Cluster (Open)
   Star Cluster (Globular)
   Galaxy

Horsehead Nebula (IC 434)

Object:

Location   Constellation of Orion, at a declination of -2 degrees

Distance   1600 light years

Size   2 light years (Horsehead itself)

Description   The iconic Horsehead Nebula is the most recognized dark nebula in the sky because of the shape of its swirling cloud of dark dust and gases, which bears resemblance to a horse's head when viewed from Earth.   The nebula lies in a field of rich objects which, on their own, would each be the star of the show. But here the Horsehead steals the attention from all of them.

The dark cloud of IC 434 is a region in the Orion Molecular Cloud Complex where star formation is taking place. One newly formed star is visible on the horse's brow. The nebula spans 5 arc-minutes in our apparent view. This corresponds to a width of 2 light years.

The nebula's reddish glow is from hydrogen gas behind the nebula, ionized by the nearby bright naked-eye star cluster Sigma Orionis just above the Horsehead. Magnetic fields steer the gases into streams, shown in the image as vertical streaks in the background glow.

The other wonderful objects in the same view, from left to right, are:
- At the left image edge, the beautiful textured red glow of IC 432
  (together with the blue haze of vdB 51).
- Just to the right of IC 432, the bright star Alnitak, a triple star whose main star is 20x the diameter of our sun.
- Just below Alnitak, the fascinating dust lane structure of the Flame Nebula (NGC 2024).
- Just to the right of the Flame, the colorful reflection nebula NGC 2023 (also known as LBN 954). Like the Horsehead, this nebula is 4 light years across.
- The blue glow of IC 435 at the bottom center of the image.

Image Acquisition:

Date   Oct 20, 2023

Exp. Time   2 hours

Telescope   Celestron RASA 11" V2
Camera   ZWO ASI6200MM Pro
Mount   Astro-Physics Mach1 GTO
Filters   Broadband - Red, Green & Blue
Narrowband - HII - A hydrogen filter was used to bring out the HII gas which is backlighting the Horsehead profile

Closeup of Horsehead:

"To the Lord your God belong the heavens, even the highest heavens."
- Deuteronomy 10:14

Object Types Included

Nebula (Dark)
X   Nebula (Emission)
Nebula (Molecular Cloud)
Nebula (PN)
Nebula (Reflection)
Nebula (SNR)
Star Cluster (Open)
Star Cluster (Globular)
Galaxy

Orion Nebula (M42)

Object:

Location   Constellation of Orion, at a declination of -5 degrees

Distance   1600 light years

Size   30 light years

Description   This iconic object is the brightest nebula in the northern hemisphere and one of the only ones visible to the naked eye, seen as the "sword" hanging beneath Orion's Belt. I think it is the most spectacular deep eye object to see from my backyard with a telescope.

At only 1600 light years away, it is one of the closest regions of massive star formation to earth. The red tint is due to large amounts of hydrogen gas. Capturing this true color image did not require narrowband filters because of the brightness of the gas colors. This beautiful object is both an emission nebula and a reflection nebula, although the red hydrogen emission is dominant in the image.

At the center of M42 is the Trapezium, which are the four stars tightly located in the center of the white portion of the image. A closeup view of the Trapezium is shown below. Each of the 4 stars also has companions, but they are small and so close to their main stars that they cannot be seen in the image below.

Besides the main attraction of M42 in the bottom half of the image on the opposite page, numerous other objects are also seen, described here moving up from M42:

- The bright circular area surrounding the bright star just left of image center is known as M43 or De Mairan's Nebula.
- The region immediately above M43 is not designated but is very interesting, with bright stars and dark dust clouds
- Near the top of the image is the Running Man Nebula (NGC 1977 or Sh2-279).

Image Acquisition:

Date   Oct 19, 2023

Exp. Time   1 hours

Telescope   Celestron RASA 11" V2
Camera   ZWO ASI6200MM Pro
Mount   Astro-Physics Mach1 GTO
Filters   Broadband - Lum, Red, Green & Blue

Closeup of Trapezium:

"He who made the Pleiades and Orion, who turns darkness into dawn and darkens day into night, who calls for the waters of the sea and pours them out over the face of the land - the Lord is his name." - Amos 5:8

Object Types Included

Nebula (Dark)
X   Nebula (Emission)
Nebula (Molecular Cloud)
Nebula (PN)
Nebula (Reflection)
Nebula (SNR)
Star Cluster (Open)
Star Cluster (Globular)
Galaxy

Pelican Nebula (IC 5070)

Object:

Location   Constellation of Cygnus, at a declination of +44 degrees

Distance   1800 light years

Size   45 light years

Description   This object, nicknamed the Pelican Nebula, is a beautiful emission nebula together with dark and billowing dust clouds. The dark clouds of this nebula conceal a large region of active star formation.

The Pelican is shown in profile, looking left. Some imagination is required to see it.

The highlight of the object is Herbig-Haro 555, a pair of bi-polar gas jets at the tip of the long trunk of towering dark dust extending to the left. This indicates the presence of a unseen protostar - a star birthed from a cold collapsing cloud of hydrogen gas. The first astronomers to study these Herbig-Haro objects in detail were George Herbig and Guillermo Haro, after whom they have been named.   A closeup of Herbig-Haro 555 is shown below.

Image Acquisition:

Date   Oct 17, 2023

Exp. Time   3.5 hours

Telescope   Celestron RASA 11" V2
Camera   ZWO ASI6200MM Pro
Mount   Astro-Physics Mach1 GTO
Filters   Broadband (for stars) - Red, Green & Blue
Narrowband (for nebula) - HII, OIII & SII
The narrowband data was mapped as follows - SII to red, HII to green & OIII to blue (SHO palette).

Closeup of Herbig-Haro 555:

"God's promises are like the stars; the darker the night the brighter they shine."
– David Nicholas

Object Types Included

    Nebula (Dark)
    Nebula (Emission)
    Nebula (Molecular Cloud)
X   Nebula (PN)
    Nebula (Reflection)
    Nebula (SNR)
    Star Cluster (Open)
    Star Cluster (Globular)
    Galaxy

Planetary Nebulae

Object:

Location   Various

Distance   3000 light years (typical)

Size   2 light years (typical)

Description   A few stars die spectacular deaths, in the form of a supernova explosion. But most die more peaceful deaths showcased by a planetary nebula (PN). A planetary nebula is the signature of a dying star. There are like snowflakes - no 2 are alike.

Planetary nebula emanate from intermediate sized stars, like our sun. These stars shed and illuminate their outer layers of gas near the end of their life. The progenitor star that remains, which is usually a small central bluish star, is destined to become a white dwarf. The sun will also die someday, long after all of us are gone.

The largest PN seen from earth is NGC 7293, seen on page 12 of this book. The 8 PN on the opposite page are some of best of the rest. All of these PN are found in our Milky Way Galaxy, realitvely close to earth. These 8 PN are found in different parts of the sky - I have superimposed them here on the same page.

Why are PN all so different? Their appearance is determined by the properties of their central star. Here are a few of the important factors, referencing the PN names on the opposite page:

- PN that have a single central star will have a spherical PN, as seen in M97. Most PN have
  binary (2) stars as their source, or even more, which leads eccentric orbits and more unusual
  PN shapes as seen in NGC 6302 and G100.4+4.6.
- Our view perspective of the PN is very important. Many PN are shaped like a hourglass.
  If this hourglass is viewed from the side, a "figure 8" shape is seen as in M76. If this
  hourglass is viewed from the end, an oval shape is seen as in M57.
- The chemical composition of the star, and the resulting expelled gas, determines the color.
  Hydrogen is red, often seen on the outer edges of PN. Oxygen is blue, often seen more
  towards the central region.
- Some PN have extended outer halos, like M57, NGC 6543, and M27. This is due to gas
  being expelled early in the formation process.

Many other factors are also involved in the PN process - each year scientists learn more about how these wonderful objects are formed.

Image Acquisition:

Date   Various

Exp. Time   5 hours (typical)

Telescope   Celestron EdgeHD 11"
Camera   ZWO ASI294MM Pro
Mount   Astro-Physics Mach1 GTO
Filters   Broadband (for stars) - Red, Green & Blue
Narrowband (for nebula) - HII & OIII
The narrowband data was mapped as follows - HII to red, OII to green & blue (HOO palette).

"There wouldn't be a sky full of stars if we were all meant to wish on the same one."
– Frances Clark

Ring
(M57)

Cat's Eye
(NGC 6543)

Dumbbell
(M27)

G100.4+04.6

Bug
(NGC 6302)

Little Dumbbell
(M76)

Headphones
(JE1)

Owl
(M97)

X   Nebula (Emission)
    Nebula (Molecular Cloud)
    Nebula (PN)
    Nebula (Reflection)
    Nebula (SNR)
X   Star Cluster (Open)
    Star Cluster (Globular)
    Galaxy

Cone Nebula (NGC 2264)

Object:

Location Constellation of Monoceros, at a declination of +10 degrees

Distance 2400 light years

Size 30 light years

Description This image captures a series of objects located in the constellation of Monoceros at a declination of +10 degrees. The primary objects in the image, from bottom left to top right, are the Cone Nebula, the Christmas Tree open star cluster (NGC 2264), the bright multiple star system S Monocerotis, the Fox Fur Nebula, and Barnard 39.

All of these features are part of the Sh2-273 nebula, a broad emission nebula which is about 4 degrees wide. It extends slightly beyond the borders of this image.

The Christmas Tree Cluster (NGC 2264) is in the lower left corner of the image but it is hard to discern. It is an open star cluster of 600 blue young stars ionizing the reddish hydrogen nebula. In the image, the "tree" is upside down. The trunk of the tree is the bright white patch of reflected light, and the top of the tree is the star at the tip of the Cone Nebula.

The most spectacular object in the image is the Cone Nebula, at the bottom left of the image. A closeup is shown below. Similar to the Pillars of Creation, the Cone Nebula is a 7-light-year-long gaseous star formation pillar silhouetted against glowing red gas. Stellar radiation from the Christmas Tree cluster is creating the cone shape as it impacts the dust. As a rule, pillars such as this one point the way to star clusters.

Image Acquisition:

Date Feb 26-27, 2023

Exp. Time 4 hours

Telescope Celestron RASA 11" V2
Camera ZWO ASI6200MM Pro
Mount Astro-Physics Mach1 GTO
Filters Broadband - Red, Green & Blue
Narrowband - HII - A hydrogen filter was used to bring out the HII gas

Closeup of Cone Nebula:

"I've loved the stars too fondly to be fearful of the night."
– Galileo Galilei

Distance from Earth (light years)

1   10   100   1k   10k   100k   1m   10m   100m   1b

Object Types Included
   Nebula (Dark)
X Nebula (Emission)
   Nebula (Molecular Cloud)
   Nebula (PN)
   Nebula (Reflection)
   Nebula (SNR)
   Star Cluster (Open)
   Star Cluster (Globular)
   Galaxy

Elephant Trunk Nebula (IC 1396)

Object:

Location   Constellation of Cepheus, at a declination of +57 degrees

Distance   3000 light years

Size   20 light years (trunk itself)

Description   This image is an emission region centered upon IC 1396, an open star cluster. The most spectacular object here is IC 1396a, the "Elephant Trunk" bright rimmed dark globule.

The trunk is over 20 light-years long. At the center of the end of the trunk are two young T Tauri stars. These stars have carved out a cavity in the globule from their powerful stellar winds. Just below and right of these stars is vdB 142, a small blue reflection nebula.

A closeup of the trunk is shown below.

Although the Elephant Trunk is the highlight here, I also enjoyed these 2 other aspects of this image:
- The many dark larger nebulae (many of them Barnard nebula) silhouetted against blue emission signal throughout the center of the nebula.
- The tight grouping of 3 bright blue stars at the center of the image. This is the star complex HD 206267, which is the center of the open star cluster IC 1396, also known as Trumpler 37. It is also the main source of excitation for the surrounding nebula.

Image Acquisition:

Date   July 28, 2023

Exp. Time   6 hours

Telescope   Celestron RASA 11" V2
Camera   ZWO ASI6200MM Pro
Mount   Astro-Physics Mach1 GTO
Filters   Broadband (for stars) - Red, Green & Blue
Narrowband (for nebula) - HII, OIII & SII
The narrowband data was mapped as follows - SII to red, HII to green & OIII to blue (SHO palette).

Closeup of Elephant Trunk:

"If the stars should appear one night in a thousand years, how would men believe and adore; and preserve for many generations the remembrance of the city of God which had been shown! But every night come out these envoys of beauty, and light the universe with their admonishing smile." – Ralph Waldo Emerson

Object Types Included

Nebula (Dark)
X   Nebula (Emission)
Nebula (Molecular Cloud)
Nebula (PN)
Nebula (Reflection)
Nebula (SNR)
Star Cluster (Open)
Star Cluster (Globular)
Galaxy

Question Mark Nebula (NGC 7822)

Object:

Location   Constellation of Cepheus, at a declination of +67 degrees

Distance   3000 light years

Size   160 light years

Description   This object is a large emission nebula that spans about 3 degrees in our apparent view, which corresponds to an actual width of 160 light years. The complex is sometimes referred to as Sh2-171 and Cederblad 214, although the true boundaries of these objects are slightly different.

Many "elephant trunks" - cosmic pillars of cold gas and dark dust – lie within this nebula. All of these trunks are pointing towards the Berkeley 59 star cluster, which is located at image center. The powerful winds and radiation of this star cluster sculpt and erode the dense pillar shapes. See below for a closeup of the central region.

The brightest stars in the image are in the foreground and do not lie within NGC 7822. Also in the foreground is one of my favorite parts of the image, the wide horizontal band across the image of distinctly shaped dark dust clouds. These clouds seem to floating by, obscuring but not completely blocking the beautiful bright emission background.

Finally, check out the interesting cloud shape towards center-left bottom. From this orientation (true north up), it looks a bit like the Super Bowl trophy (see closeup below). I could not find any documentation of this object, except that it is called simply IRAS 00029+6546 -- Cloud.

Image Acquisition:

Date   Aug 17-18, 2023

Exp. Time   6 hours

Telescope   Celestron RASA 11" V2
Camera   ZWO ASI6200MM Pro
Mount   Astro-Physics Mach1 GTO
Filters   Broadband (for stars) - Red, Green & Blue
Narrowband (for nebula) - HII, OIII & SII
The narrowband data was mapped as follows - SII to red, HII to green & OIII to blue (SHO palette).

Closeup of: Central Region

Odd Cloud Shape

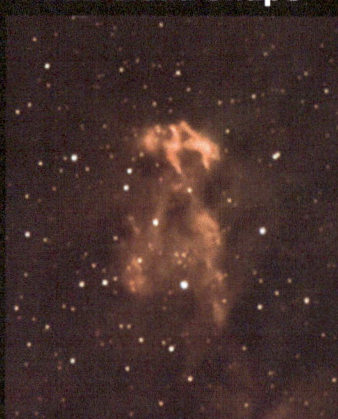

"We are all of us stars, and we deserve to twinkle."
– Marilyn Monroe

Object Types Included

Nebula (Dark)
Nebula (Emission)
Nebula (Molecular Cloud)
Nebula (PN)
Nebula (Reflection)
X   Nebula (SNR)
Star Cluster (Open)
Star Cluster (Globular)
Galaxy

Spaghetti Nebula (Sh2-240)

Object:

Location   Constellations of Auriga and Taurus, at a declination of +28 degrees

Distance   3000 light years

Size   150 light years

Description   This object is a huge faint supernova remnant (SNR) . Most stars meet their end in the form of a planetary nebula, but some explode violently in one of the most powerful explosions that the universe has seen. Scientists believe that this star exploded about 40,000 years ago. Shortly after its explosion, the supernova was so bright that it could likely have been seen in our daytime sky. Now, all we see are the faint gas arcs that keep expanding in all directions.

This image, in the HOO palette, captures HII filaments in red and OIII filaments in cyan, with hydrogen the more dominant gas. Many of the other SNRs I have imaged have a more equal balance of hydrogen and oxygen. The IC 443 nebula (Jellyfish Nebula, p. 50) is one SNR which has a similar HII dominance to SH2-240. Another similarity to the smaller IC 443 nebula is that the spherical nature of the SNR is still mostly intact in both nebula.

Numerous irregular dark foreground clouds somewhat obscure the nebula, particularly at the lower center and in the upper right quadrant. Some extended lobes are seen, which could be due to various factors such as non-uniform density of the interstellar medium, or non-symmetric expulsions from the explosion. The most significant lobes are polar-opposite up and down. It is interesting that, in contrast to the rest of the nebula, the outermost extent of these lobes appear to be dominated more by oxygen than by hydrogen.

A pulsar near the center of this nebula, not visible in this image, is the progenitor (source) star.

This object is also known as Simeis 147 (S147). This identifier was established in a paper by Gaze and Shajn in a Crimean journal in the 1950's. The name Simeis originates with an older astronomy facility operated by the Crimean Astrophysical Observatory in the city of Simeiz (in current day Ukraine). For some reason, the city spelling ends in "z" while the observatory ends in "s". The only other popular Simeis object is Simeis 57 (The Propeller Nebula, sometimes referred to as DWB 111).

Image Acquisition:

Date   Jan 14-16, 2023

Exp. Time   8 hours

Telescope   Celestron RASA 11" V2
Camera   ZWO ASI6200MM Pro
Mount   Astro-Physics Mach1 GTO
Filters   Broadband (for stars) - Red, Green & Blue
Narrowband (for nebula) - HII & OIII
The narrowband data was mapped as follows - HII to red, OIII to both green & blue (HOO palette).

"The real friends of the space voyager are the stars. Their friendly, familiar patterns are constant companions, unchanging, out there." - James Lovell (astronaut)

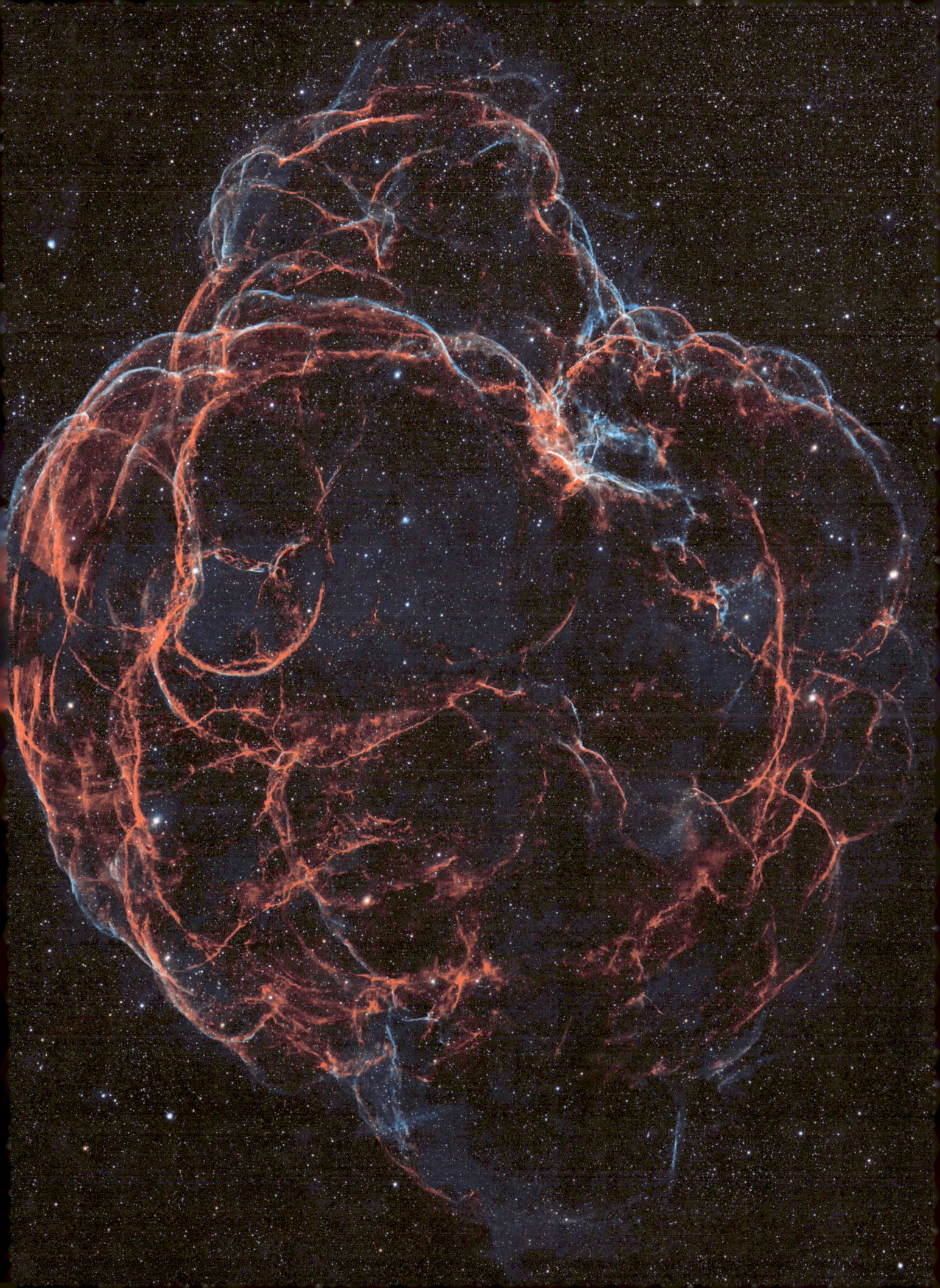

Object Types Included

Nebula (Dark)
X   Nebula (Emission)
Nebula (Molecular Cloud)
Nebula (PN)
Nebula (Reflection)
Nebula (SNR)
Star Cluster (Open)
Star Cluster (Globular)
Galaxy

Squid & Flying Bat Nebula (OU4)

Object:

Location   Constellation of Cepheus, at a declination of +60 degrees

Distance   3400 light years

Size   60 light years

Description   This image captures a pair of emission nebulae -- Sh2-129 (The Flying Bat Nebula) shown in the red ionized hydrogen emissions, and the very faint OU-4 (The Giant Squid Nebula) shown in the cyan ionized oxygen emissions.

OU-4 was recently discovered in 2011. It is very faint but very large in the sky, spanning the equivalent of 2 full moons. This nebula is estimated to be 100,000 years old based upon its outflow and size. It is about 60 light years long.

The structure and size of OU-4 is unusual. Although we do not know its exact origin, it seems logical that it has formed from ionized oxygen emissions emanating from the bright triple star system (HR 8119) which it is centered upon. One possibility is that the outflow is produced from an eruptive outburst caused by mass accretion in a binary system. This event is called an intermediate–luminosity optical transient (ILOT).

So, is this a planetary nebula? Some people, dating back many years, consider PN to simply be gas ejected from a star, which this is. But a PN is now defined as a shell of ionized gas ejected from a red giant star late in its life. The shell mechanism seems more complicated than that here. More importantly, the typical size of a PN is about 1 to 3 light years or less. This one extends 50 light years. My belief is that this is not a PN, but I will leave that debate to the scientists.

I like the complex multi-polar structure of this nebula. Bow shocks are seen in multiple places, particularly the bright arc at the bottom of the nebula. The radius of all of these bowshocks point back to the triple star system. The nebula looks to be a continuous surface bubble except in two spots about 1/4 of the way up from the bottom of the nebula, where the outflow appears to have broken through the bubble.

Image Acquisition:

Date   Oct 15-16, 2023

Exp. Time   3.5 hours

Telescope   Celestron RASA 11" V2
Camera   ZWO ASI6200MM Pro
Mount   Astro-Physics Mach1 GTO
Filters   Broadband (for stars) - Red, Green & Blue
Narrowband (for nebula) - HII & OIII
The narrowband data was mapped as follows - HII to red, OIII to both green & blue (HOO palette)..

"I watched the night sky with its countless stars and its moon, and I wondered about the universe and all that had been created, why the stars and the moon rose at night and the sun in the day, how vast it must be, how I could never understand the infinite measure of its size." - Patrick Carman

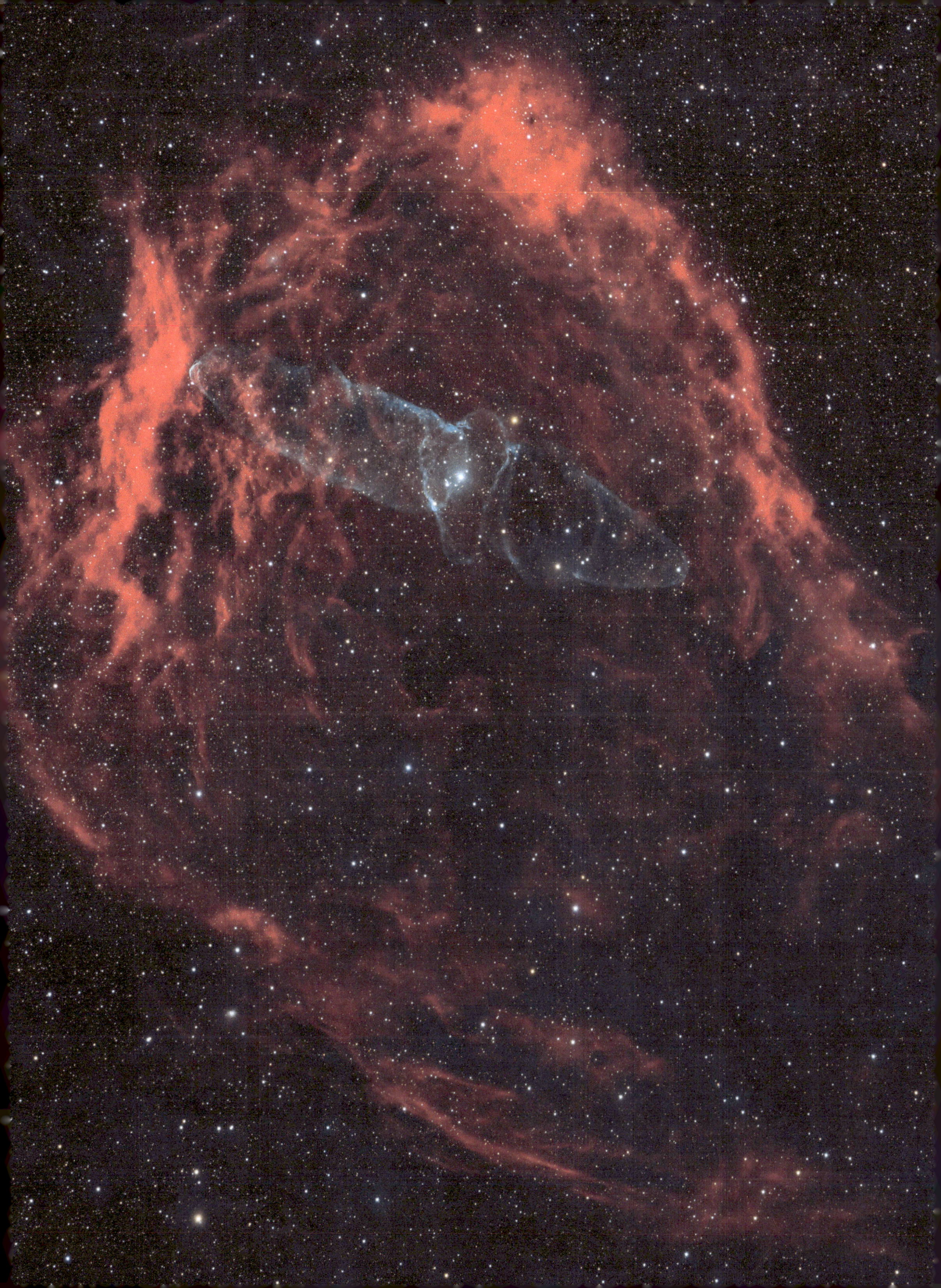

Object Types Included

Nebula (Dark)
X Nebula (Emission)
Nebula (Molecular Cloud)
Nebula (PN)
Nebula (Reflection)
Nebula (SNR)
Star Cluster (Open)
Star Cluster (Globular)
Galaxy

Seagull Nebula

Object:

Location — Constellations of Monoceros and Canis Major, at a declination of -11 degrees

Distance — 3600 light years

Size — 250 light years

Description — This wonderful object, nicknamed the Seagull Nebula, consists of many amazing objects:

- The beautiful Seagull head (Gum 1 / SH2-292)
- The Seagull wings (Gum2 / IC 2177)
- The large star cluster at the leading edge of the left wing (NGC 2335)
- The small cluster at the leading edge of the right wing (NGC 2327)
- The nebula at the tip of the right wing (Gum 3 / Sh2-297)
- The cluster at the left claw (NGC 2343), and
- The two areas of nebulosity at the top right (Sh2-293 and Sh2-295).

Many of these objects are also classified as vdB reflection nebulae objects.

The brilliance of the nebula is accentuated by the dense Milky Way star field which surrounds it.

I also like the many small hidden gems which are seen here upon close inspection. These include the colorful Kohoutek 1-8 planetary nebula (just above and right of the seagull head), the blue arc of the vdB 95 reflection nebula (at the right claw), and the molecular cloud around star HD 54359 that is a YSO factory, looking like an alien spaceship taking off (at lower left, below NGC 2335). Closeups of these objects are shown below.

Image Acquisition:

Date — Jan 26-27, 2022

Exp. Time — 7 hours

Telescope — Celestron RASA 11" V2
Camera — ZWO ASI6200MM Pro
Mount — Astro-Physics Mach1 GTO
Filters — Broadband (for stars) - Red, Green & Blue
Narrowband (for nebula) - HII, OIII & SII
The narrowband data was mapped as follows - SII to red, HII to green & OIII to blue (SHO palette).

Closeups of:

Kohoutek 1-8 PN

vdB 95

Nebula around HD 54359

"Observe the stars, millions of them, twinkling in the night sky, all with a message of unity, part of the very nature of God." - Sai Baba

Object Types Included

Nebula (Dark)
X Nebula (Emission)
Nebula (Molecular Cloud)
Nebula (PN)
Nebula (Reflection)
Nebula (SNR)
Star Cluster (Open)
Star Cluster (Globular)
Galaxy

Lagoon Nebula (M8)

Object:

Location   Constellation of Sagittarius, at a declination of -24 degrees

Distance   4000 light years (M8)

Size   110 light years (M8)

Description   The Lagoon Nebula has an apparent size to us of about three times the size of the full Moon.

I find it interesting that visual observers struggle to understand why this object has become known as the Lagoon Nebula. For us narrowband imagers, using the Hubble SHO palette, the deep blue of the central region is easy to imagine as an inviting lagoon.

A large cluster (designated NGC 6530) of 100 stars is powering the nebula. This cluster, located just above the nebula center, is beautiful in itself and would stand out if not surrounded by such a bright nebula. Dark Bok globules (small inky black winding dust lanes) are scattered everywhere throughout the nebula.   These Bok globules are collapsing protostellar clouds which birth new stars.

I like how the "water" (blue OIII signal) of the Lagoon seems to be overflowing and running out over the rim of the lake on all 4 sides.

The bright region just left of and below the center of the image contains the bright Hourglass, which can be tough to spot at first. This is not a bipolar nebula, but simply a result of overlapping dark nebulae clouds. The Hourglass is our unobstructed window into the bright interior of the emission region, as shown in the closeup below and particularly in the inverted view.

Image Acquisition:

Date   July 15 & 18, 2023

Exp. Time   4 hours

Telescope   Celestron RASA 11" V2
Camera   ZWO ASI6200MM Pro
Mount   Astro-Physics Mach1 GTO
Filters   Broadband (for stars) - Red, Green & Blue
Narrowband (for nebula) - HII, OIII & SII
The narrowband data was mapped as follows - SII to red, HII to green & OIII to blue (SHO palette).

Closeup of Central M8 Region (inverted view shown for detail):

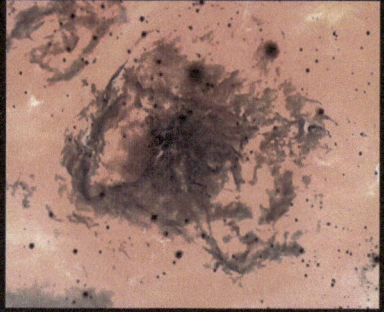

"The cosmos is all that is or ever was or ever will be. Our feeblest contemplations of the Cosmos stir us—there is a tingling in the spine, a catch in the voice, a faint sensation, as if a distant memory, or falling from a height. We know we are approaching the greatest of mysteries." - Carl Sagan, Astronomer

Object Types Included

   Nebula (Dark)
X Nebula (Emission)
   Nebula (Molecular Cloud)
   Nebula (PN)
   Nebula (Reflection)
   Nebula (SNR)
   Star Cluster (Open)
   Star Cluster (Globular)
   Galaxy

Dragons of Ara (NGC 6188)

Object:

Location   Constellation of Ara, at a declination of -49 degrees

Distance   4000 light years

Size   100 light years

Description   This emission region has numerous nicknames, including the Rim Nebula, the Shaking Hands Nebula, and my favorite DSO nickname, the Dragons of Ara. The "dragons" are fighting in the region just above and left of image center. A closeup of the dragons is shown below.

Like familiar objects such as the Eagle Nebula and the Cygnus Wall, this object features a large dust cloud that has been eroded into numerous billowing pillars by stellar winds, silhouetted against the emission nebula. The nebula is powered by the open star cluster NGC 6193, located just above the dragons.

Seen towards the bottom right of the image is the beautiful Gum 52 nebula. A closeup of the PN is shown below. This object, also known as RCW 107, is an emission nebula located in the constellation of Norma. This nebula spans 7 arc-minutes in our apparent view, which corresponds to an actual diameter of 7 light years. The bright white central star is clearly visible. This is a rare star type (type O6.5f, one of only 5 known) and is a massive hot bi-polar star. The resulting nebula is an emission region and not a planetary nebula. I love the intricate details of the nebula, as well as the bright areas at each end.

In the full image, note the faint blue gas halo seen much further away from the central star of Gum 52 than the PN.

Image Acquisition:

Date   April 22, 2023

Exp. Time   4 hours

Telescope   Celestron RASA 11" V2
Camera   ZWO ASI6200MM Pro
Mount   Astro-Physics Mach1 GTO
Filters   Broadband (for stars) - Red, Green & Blue
Narrowband (for nebula) - HII, OIII & SII
The narrowband data was mapped as follows - SII to red, HII to green & OIII to blue (SHO palette).

Closeups -   Fighting Dragons: Gum 52:

"Astronomy taught us our insignificance in Nature." - Ralph Waldo Emerson, Poet

Object Types Included

- Nebula (Dark)
- X Nebula (Emission)
- Nebula (Molecular Cloud)
- Nebula (PN)
- Nebula (Reflection)
- X Nebula (SNR)
- Star Cluster (Open)
- Star Cluster (Globular)
- Galaxy

Jellyfish Nebula (IC 443)

Object:

Location   Constellation of Gemini, at a declination of +23 degrees

Distance   5000 light years

Size   85 light years

Description   This image consists of many different types of deep sky objects. At lower left, IC 443 (Sh2-248) is a large supernova remnant (SNR) and molecular cloud. Just right of image center is IC 444, a small reflection nebula. At upper right is the Sh2-249 emission nebula, along with numerous dark nebulae.

IC 443, the Jellyfish Nebula, is about twice the apparent width as our full moon, and is about 85 light-years wide. The image shows the hydrogen, oxygen, and sulfur gas fronts which resulted from the violent star explosion, estimated to have occurred about 10,000 years ago. The remnants of the now dense neutron progenitor star which originated the explosion are located in the center of the SNR. Detail of the brightest portion of the SNR is shown below.

The morphology of the IC 443 SNR has been long studied and debated. The image shows that it is complex. A 2006 paper concluded that it mainly consists of 2 expanding shells of different diameters. The left shell is smaller because it is being constrained by a neutral H1 cloud. At the center of the SNR, in the foreground, is an obscuring molecular cloud which darkens the central portion of the SNR.

Image Acquisition:

Date   Jan 10, 2023

Exp. Time   4 hours

Telescope   Celestron RASA 11" V2
Camera   ZWO ASI6200MM Pro
Mount   Astro-Physics Mach1 GTO
Filters   Broadband (for stars) - Red, Green & Blue
Narrowband (for nebula) - HII, OIII & SII
The narrowband data was mapped as follows - SII to red, HII to green & OIII to blue (SHO palette).

Closeup of Brightest SNR Region:

"The heavens proclaim his righteousness, and all the peoples see his glory." - Psalms 97:6

Object Types Included

Nebula (Dark)
X   Nebula (Emission)
Nebula (Molecular Cloud)
Nebula (PN)
Nebula (Reflection)
Nebula (SNR)
Star Cluster (Open)
Star Cluster (Globular)
Galaxy

Rosette Nebula (NGC 2237)

Object:

Location	Constellation of Monoceros, at a declination of +5 degrees
Distance	5000 light years
Size	150 light years

Description This object consists of a central star cluster which has blown a large cavity into the surrounding molecular cloud. The star cluster is NGC 2244 (Caldwell 50) and the surrounding large emission nebula is NGC 2237 (Caldwell 49). The object, nicknamed the Rosette Nebula, is located 5000 light years away in the constellation of Monoceros at a declination of +5 degrees. The nebula is about 1.7 degrees in width and 150 light years in diameter.

The Rosette Nebula is the most famous example of a Strömgren sphere, where a sphere of ionized hydrogen (H II) stabilizes around a young star of the spectral classes O or B. The theory for these spheres was derived by Dr. Bengt Strömgren in 1937 and later named Strömgren sphere after him.

Numerous star-producing dark Bok globules are visible in the lower right central region of the nebula. This train of globules is known as the Circus Animal Parade, which I show below in a closeup. The globules are curving in various directions, but the windswept heads of all of them point back to the cluster center.

Less developed globules, looking like puffy clouds, are seen in the full image on the opposite side of the central star cluster, towards the left. Also seen are bright whitish wisps of nebula arcs, seen most prominently just above and to the left of the central star cluster.

In 2019, Oklahoma made the Rosette Nebula its official state astronomical object. I am not exactly sure why states need official astronomy objects.

Image Acquisition:

Date	Jan 26-27, 2023
Exp. Time	4 hours
Telescope	Celestron RASA 11" V2
Camera	ZWO ASI6200MM Pro
Mount	Astro-Physics Mach1 GTO
Filters	Broadband (for stars) - Red, Green & Blue
	Narrowband (for nebula) - HII, OIII & SII
	The narrowband data was mapped as follows - SII to red, HII to green & OIII to blue (SHO palette).

Closeup of Circus Animal Parade:

"The stars are the jewels of the night, and perchance surpass anything which day has to show."
– Henry David Thoreau

X Nebula (Dark)
Nebula (Emission)
Nebula (Molecular Cloud)
Nebula (PN)
Nebula (Reflection)
Nebula (SNR)
Star Cluster (Open)
Star Cluster (Globular)
Galaxy

Crescent Nebula (NGC 6888)

Object:

Location Constellation of Cygnus, at a declination of +38 degrees

Distance 5000 light years

Size 27 light years

Description This object is a beautiful nebula consisting of an expanding shell of gas about 27 light-years across. The shell is being blown out by winds from its central, bright, massive star. The central star is classified as a Wolf-Rayet star, so the resulting nebula is called a Wolf-Rayet nebula. Such stars suffer a high rate of mass loss, which often results in spectacular nebulae. Other Wolf-Rayet nebulae in this book include Sh2-308 (Gourd, p. 56) and NGC 2359 (Thor's Helmet, p. 84).

The progenitor (source) star is designated WR136 and is the bright star at the center of the nebula. The star is shedding its outer envelope in a strong stellar wind, ejecting the equivalent of the Sun's mass every 10,000 years. The nebula's complex structures are the result of this strong wind interacting with material ejected in earlier phases.

This image is bi-color, using light from hydrogen (HII) and oxygen (OIII) atoms in the wind-blown nebula and then adding RGB stars. The oxygen atoms produce the blue-green hue that is predominantly on the outside of the nebula, while the hydrogen atoms produce the reddish hue that is predominantly on the inside. For some reason this structure is reversed from the typical PN structure, which has hydrogen on the outside and oxygen on the inside.

My favorite parts of this object are the billowing arcs of oxygen gas fronts seen at both the top and bottom of the nebula. At the top they look like a wispy, spiffy cap, while at the bottom they look like puffy mammatus clouds.

The surrounding seas of HII nebulosity contains another interesting object. The Soap Bubble PN (also known as PN G75.5+1.7) is faintly seen between the Crescent Nebula and the bottom of the page. This PN, discovered just 15 years ago, is 4000 light years away and is 5 light-years in diameter. I have attached a closeup of both the Crescent and the Soap Bubble below.

Image Acquisition:

Date Sep 18, 2023

Exp. Time 3 hours

Telescope Celestron RASA 11" V2
Camera ZWO ASI6200MM Pro
Mount Astro-Physics Mach1 GTO
Filters Broadband (for stars) - Red, Green & Blue
Narrowband (for nebula) - HII & OIII
The narrowband data was mapped as follows - HII to red, OIII to both green & blue (HOO palette).

Closeups -

Crescent Nebula:

Soap Bubble

""For my part, I know nothing with any certainty but the sight of the stars makes me dream."

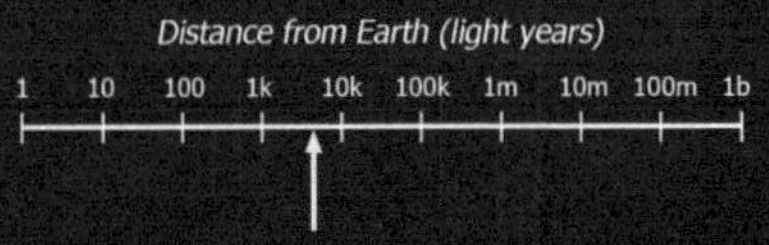

Object Types Included

Nebula (Dark)
X   Nebula (Emission)
Nebula (Molecular Cloud)
Nebula (PN)
Nebula (Reflection)
Nebula (SNR)
Star Cluster (Open)
Star Cluster (Globular)
Galaxy

Gourd Nebula (Sh2-308)

Object:

Location   Constellation of Canis Major, at a declination of -24 degrees

Distance   5000 light years

Size   60 light years

Description   This object, also known as the Dolphin's Head Nebula, is another Wolf-Rayet emission nebula. Wolf-Rayet emission nebulae are some of the most beautiful objects in the universe.

The nebula is being blown out by fast winds from a hot huge 6.8 magnitude Wolf-Rayet star (HD 50896), the bright blue star near the center of the nebula. Wolf-Rayet stars have over 20 times the mass of our Sun and are thought to be in a brief, pre-supernova phase. The gas in this nebula is primarily OIII gas.

So many things grab my attention here:

- Although the nebula is roughly spherical, it is elongated along an axis that runs from bottom left to top right of the image. This is likely due to bi-polar gas expulsion from the opposite stellar poles. The bulge at top right appears to be on the verge of breaking out. Several other minor axes of gas expulsion are also seen here, particularly the small bumps at left and right.

- My favorite part of this object, aside from the overall shape, is how the OIII nebula is slightly larger in diameter than the HII nebula. This is seen in the image where the outer portion of the white nebula surface sits slightly inside the outer portion of the cyan oxygen surface. This is most obvious on the right side of the nebula.

- The red pattern of HII towards the top of the nebula is tough to understand. Is it part of the nebula, or simply part of the background HII in the sky? It seems like it is part of the background to me.

- The bright 3.9 magnitude orange/white supergiant star towards the bottom of the nebula is Omicron Canis Majoris (HD 50877). It is about 2500 light years away, about halfway between us and the nebula. This cool (4000 K) star is about 8 times as massive as our Sun, 280 times its diameter, and shines with 16,000 times its luminosity. Its low temperature is responsible for the orange color, as compared to the blue color of hot stars.

- The faint small dim whitish perfectly circular object at the left edge of the nebula (9 o'clock) is the planetary nebula PN G234.9-09.7. Little information is available for this PN. If it is the typical PN diameter of 2 light years, this PN would be about 2500 light years away, about the same distance away as the bright orange star.

Image Acquisition:

Date   Nov 28-29, 2022

Exp. Time   4.5 hours

Telescope   Celestron RASA 11" V2
Camera   ZWO ASI6200MM Pro
Mount   Astro-Physics Mach1 GTO
Filters   Broadband (for stars) - Red, Green & Blue
Narrowband (for nebula) - HII & OIII
The narrowband data was mapped as follows - HII to red, OIII to both green & blue (HOO palette).

"Without the dark, we'd never see the stars. There also would be no use for the moon if there was never a night."
- Tessa Emily Hall

Object Types Included

  Nebula (Dark)
X Nebula (Emission)
  Nebula (Molecular Cloud)
  Nebula (PN)
  Nebula (Reflection)
  Nebula (SNR)
  Star Cluster (Open)
  Star Cluster (Globular)
  Galaxy

Eagle Nebula (M16)

Object:

Location Constellation of Serpens, at a declination of -14 degrees

Distance 5500 light years

Size 60 light years

Description This is the famous M16, nicknamed the Eagle Nebula, consisting of a central star cluster (NGC 6611) and a surrounding emission nebula (IC 4703).

The large star cluster at the heart of the Eagle contains approximately 8000 stars and is the source for the ionization of the surrounding gas clouds. The darker areas of dense gas are believed to be the sites of new star formation. Two such areas include the "Pillars of Creation", at the center of the image, and the "Stellar Spire" above and left of the pillars.

The Pillars of Creation formation was made famous in a 1995 Hubble telescope photo and has become the most recognizable Hubble image. I have attached a closeup of this region below. The pillars are being eroded by the light from nearby stars that have recently formed. The small dark areas in the image are believed to be protostars (Bok globules), containing dense cosmic dust and gas from which star formation may take place.

This is my favorite deep sky object object. I still remember my first grainy view of it on my telescope computer monitor while first imaging it outside under the stars long ago - it took my breath away then and I am still memorized by it now. I love how the fantastic pillars are sculpted by the stellar winds coming from the star cluster, how they are located at the very heart of the eagle, and how they are silhouetted against the deep blue colors of the central nebula region.

Image Acquisition:

Date April 28-29, 2023

Exp. Time 4 hours

Telescope Celestron RASA 11" V2
Camera ZWO ASI6200MM Pro
Mount Astro-Physics Mach1 GTO
Filters Broadband (for stars) - Red, Green & Blue
Narrowband (for nebula) - HII, OIII & SII
The narrowband data was mapped as follows - SII to red, HII to green & OIII to blue (SHO palette).

Closeup of Pillars of Creation:

"In the beginning God created the heavens and the earth." - Genesis 1:1

Object Types Included

   Nebula (Dark)
X Nebula (Emission)
   Nebula (Molecular Cloud)
   Nebula (PN)
   Nebula (Reflection)
   Nebula (SNR)
   Star Cluster (Open)
   Star Cluster (Globular)
   Galaxy

Cat's Paw Nebula (NGC 6334)

Object:

Location   Constellation of Scorpius, at a declination of -36 degrees

Distance   5500 light years

Size   50 light years

Description   This object, also known as Sh2-8 and the Cat's Paw Nebula, is about 1/2 degree in diameter, which corresponds to a width of about 50 light years. It is about the same size in the sky as our full moon.

This narrowband image is dominated by an orange/yellow hue from a combination of the strong hydrogen (HII) and sulfer (SII) signals, relative to the weak overall oxygen (OIII) signal.   The multihued bottom left lobe is due to the stronger presence of oxygen in that lobe. I like the dark dust lanes throughout the nebula which are silhouetted against the emission signal.

This region of the sky is one of the most prolific star forming regions in the Milky Way. Scientists cannot definitively identify the cause for such intense star formation in this nebula, but they have ruled out the usual triggers of a supernova or colliding galaxies.

Image Acquisition:

Date   Aug 15-16, 2023

Exp. Time   4 hours

Telescope   Celestron RASA 11" V2
Camera   ZWO ASI6200MM Pro
Mount   Astro-Physics Mach1 GTO
Filters   Broadband (for stars) - Red, Green & Blue
Narrowband (for nebula) - HII, OIII & SII
The narrowband data was mapped as follows - SII to red, HII to green & OIII to blue (SHO palette).

"If the stars should appear one night in a thousand years, how would men believe and adore; and preserve for many generations the remembrance of the city of God which had been shown! But every night come out these envoys of beauty, and light the universe with their admonishing smile." – Ralph Waldo Emerson

Object Types Included

Nebula (Dark)
Nebula (Emission)
Nebula (Molecular Cloud)
Nebula (PN)
Nebula (Reflection)
X   Nebula (SNR)
Star Cluster (Open)
Star Cluster (Globular)
Galaxy

Crab Nebula (M1)

Object:

Location   Constellation of Taurus, at a declination of +22 degrees

Distance   6500 light years

Size   13 light years

Description   This showcase object is a mag 8.4 supernova remnant located 6500 light years away in the constellation of Taurus. Ii is about 7 arc-minutes in diameter, which corresponds to an actual diameter of 13 light years.

M1 is one of the few deep sky objects whose birth was witnessed by man. Chinese astronomers watching the sky on July 4, 1054, noted the appearance of a new star in the night sky. The bright source was visible during the daytime for 23 days, shining six times as brightly as Venus. Astronomers were able to see it in the night sky with the naked eye for almost two years. Other observations were recorded by Japanese, Arabic, and Native American stargazers. This "new star" was actually the death of an old star in a violent supernova explosion.

The object that you see here is not that bright star, but the aftermath of the explosion in the form of a "supernova remnant" (SNR) nebula. This faint nebula was discovered 667 years after the explosion, in 1721 by the British astronomer John Bevis.

This object is designated as M1, the first object in Charles Messier's 250 year old catalog of deep sky objects, the Messier Catalog. Messier was a world-renowned comet hunter.   King Louis XV of France nicknamed him the "Ferret of Comets".   He is most famous now for his deep sky catalog. Ironically, at the time, these deep sky objects where documented not because they were beautiful, but because he grew tired of mistaking these objects for comets and documented them so that he would no longer waste his time on them during his comet hunts.

The Crab Nebula nickname is due to its resemblance to a crab's claw (not the full-body image of a crab) in an early sketch made in 1855. But in this photographic image, it doesn't really resemble a crab claw.

The fine tendrils seen throughout the nebula are ionized hydrogen, oxygen and sulfur gas fronts, ejected in the supernova explosion. The different colors are reflective of the variation in the strengths of these gas shock waves. The nebula continues to rapidly expand outward, changing the details of its appearance slightly each year.

Image Acquisition:

Date   Feb 11-14, 2022

Exp. Time   12 hours

Telescope   Celestron EdgeHD 11"
Camera   ZWO ASI294MM Pro
Mount   Astro-Physics Mach1 GTO
Filters   Broadband (for stars) - Red, Green & Blue
Narrowband (for nebula) - HII, OIII & SII
The narrowband data was mapped as follows - SII to red, HII to green & OIII to blue (SHO palette).

"God made the earth by his power; he founded the world by his wisdom and stretched out the heavens by his understanding." - Jeremiah 10:12

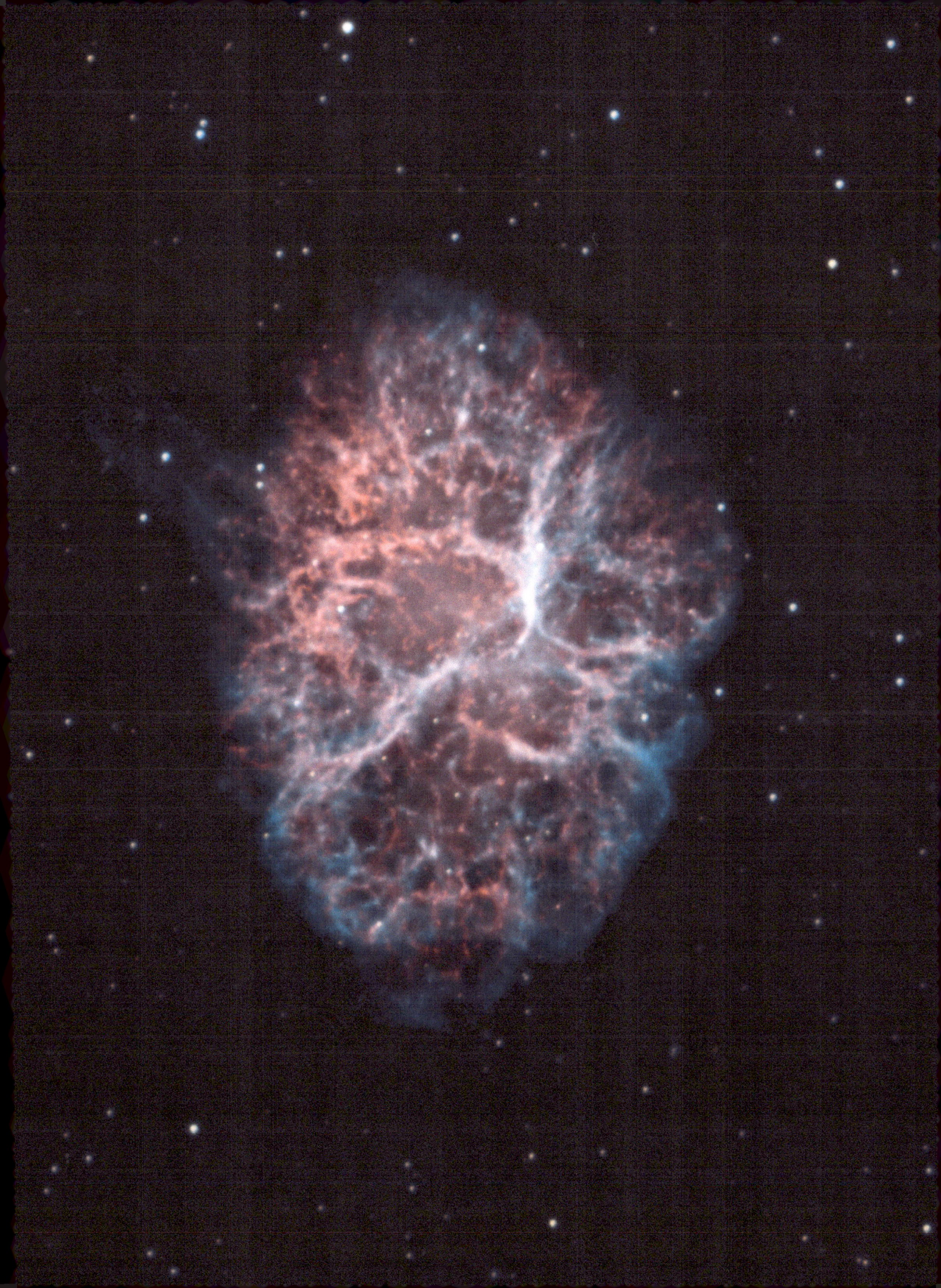

Object Types Included

Nebula (Dark)
X Nebula (Emission)
Nebula (Molecular Cloud)
Nebula (PN)
Nebula (Reflection)
Nebula (SNR)
Star Cluster (Open)
Star Cluster (Globular)
Galaxy

Soul Nebula (IC 1848)

Object:

Location   Constellation of Cassiopeia, at a declination of +60 degrees

Distance   6500 light years

Size   180 light years

Description   This is a busy complex. IC 1848 is the open star cluster that occupies the lower blue lobe in the image. Embedded in the upper blue lobe are several smaller open clusters including IC 1871 and Collinder 34. The dramatic ridge of gases running horizontally across the center of the page and separating the two regions is LBN 673.

The stellar winds of the open star clusters in each region are carving out two huge evacuated lobes. This process leaves behind large pillars of eroded dust, all pointing inwards towards the star clusters in each lobe. These pillars are very dense and have stars forming at their tips. Each pillar spans about 10 light years. The pillars exist around half of the circumference of each lobe.

One of my favorite parts of this image is the small bright nebula, Sh2-201, seen at the very top of the image. The internal dust patterns look alien-like to me. A closeup of this region is seen below.

I sometimes incorrectly think of space objects as being two dimensional because that is all we can see on a flat screen or book page, but it is interesting to imagine what these blue spherical lobes must look like in 3 dimensions. The blue regions in each lobe, the central star clusters, the surrounding dust pillars, and the wispy faint black nebulae at the top are all easier to understand for me if I visualize it as a 3-D space.

This beautiful object is unfortunately stuck with the nickname of the Soul Nebula. It doesn't really look like a soul to me, whatever that would look like. It is simply located next to the Heart Nebula. Heart and Soul, get it? I am not a big fan of nicknames, and this is a good example why. If I was going to name it, it would be the Buffalo Nebula (tilt your head to the left to see it).

Image Acquisition:

Date   Jan 8-9, 2023

Exp. Time   4 hours

Telescope   Celestron RASA 11" V2
Camera   ZWO ASI6200MM Pro
Mount   Astro-Physics Mach1 GTO
Filters   Broadband (for stars) - Red, Green & Blue
Narrowband (for nebula) - HII, OIII & SII
The narrowband data was mapped as follows - SII to red, HII to green & OIII to blue (SHO palette).

Closeup of Sh2-201:

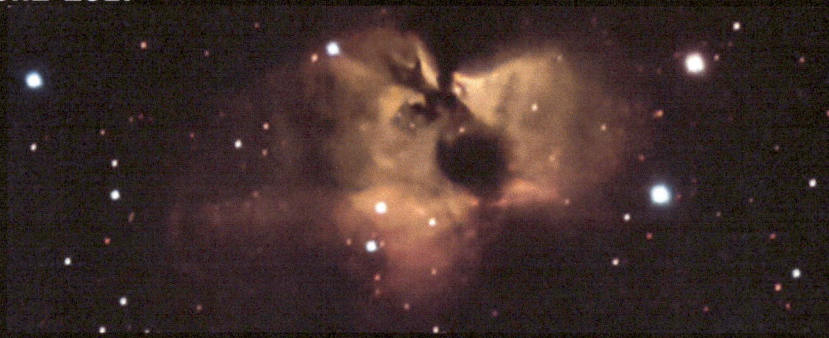

"For my confirmation, I didn't get a watch and my first pair of long pants, like most Lutheran boys. I got a telescope. My mother thought it would make the best gift." - Wernher von Braun, Rocket Engineer

Distance from Earth (light years)

1 10 100 1k 10k 100k 1m 10m 100m 1b

Object Types Included

Nebula (Dark)
X Nebula (Emission)
Nebula (Molecular Cloud)
Nebula (PN)
Nebula (Reflection)
Nebula (SNR)
Star Cluster (Open)
Star Cluster (Globular)
Galaxy

Heart Nebula (IC 1805)

Object:

Location   Constellation of Cassiopeia, at a declination of +61 degrees

Distance   7500 light years

Size   180 light years

Description   This emission nebula spans 1.5 degrees in our apparent view, which corresponds to an actual width of about 180 light years. It is about 6 times wider than the full moon in the sky.

This object is a fantastic mix of bright stars, glowing interstellar gas, and dark dust clouds. In the center of the image are the massive hot stars of a newborn star cluster known as Melotte 15. The stellar winds and radiation of these stars are sculpting the surrounding dust clouds into fantastic twisting shapes and pillars. A closeup of this amazing region is shown below.

At upper right is the colorful Fishhead Nebula (IC1795), which has interestingly shaped dust lanes cutting through the nebula.

Image Acquisition:

Date   Dec 26, 2022

Exp. Time   4 hours

Telescope   Celestron RASA 11" V2
Camera   ZWO ASI6200MM Pro
Mount   Astro-Physics Mach1 GTO
Filters   Broadband (for stars) - Red, Green & Blue
Narrowband (for nebula) - HII, OIII & SII
The narrowband data was mapped as follows - SII to red, HII to green & OIII to blue (SHO palette).

Closeup of Central Region (Melotte 15):

"Look up at the stars and not down at your feet. Try to make sense of what you see, and wonder about what makes the universe exist. Be curious." - Stephen Hawking

Nebula (Dark)
X Nebula (Emission)
Nebula (Molecular Cloud)
Nebula (PN)
Nebula (Reflection)
Nebula (SNR)
Star Cluster (Open)
Star Cluster (Globular)
Galaxy

Bubble Nebula (NGC 7635)

Object:

Location   Constellation of Cassiopeia, at a declination of +61 degrees

Distance   7800 light years

Size   7 light years

Description   The bubble is caused by a stellar wind emitted from the source star, which is the brightest star inside of the bubble. As seen in the closeup below, this source star is not centered in the bubble.

This star is a Wolf-Rayet star, which is an O-type (hot and blue-white) star that is nearing supernova stage. These stars expel a significant amount of mass in this stage, resulting in the expanding gas bubble which is contained by the molecular cloud.

The surrounding cloud has been shaped by the gas velocity into orange-tinted dust pillars which all point back to the source star. The pillars are similar to the iconic columns in the "Pillars of Creation" Eagle Nebula. As seen with the structures in the Eagle Nebula, the Bubble Nebula pillars are being illuminated by the strong ultraviolet radiation from the brilliant star inside the bubble. A bright pillar, located behind the bubble, can be seen shining through the bubble just right of the source star. Cometary knots are seen throughout the image at the tips of the pillars.

Note the shape of the bubble, which is not quite symmetric. The bubble has a bulge towards the right side. This asymmetry is likely due to polar directional outflow from the star, but could also be due to differences in the density of the surrounding gaseous material. Looking closely at the full image, you can see the faint outline of an even larger bubble which is over twice as large as the main bubble. The larger, fainter bubble also has a bit of a point on the right side, which coincides with the brightest foreground star in the image. This larger bubble was likely formed by earlier gas releases from the source star. Hints of even larger bubbles, in the form of faint arcing wave fronts, can also be seen in the image.

I like the 3-D look of the bubble, especially how the curved outer surface is highlighted by white reflected starlight. This is seen most brightly at the top left quadrant of the bubble. I also like how the bubble seems to serve as a type of magnifying glass on the background behind it.

Image Acquisition:

Date   Nov 28, 2023

Exp. Time   4 hours

Telescope   Celestron RASA 11" V2
Camera   ZWO ASI6200MM Pro
Mount   Astro-Physics Mach1 GTO
Filters   Broadband (for stars) - Red, Green & Blue
Narrowband (for nebula) - HII, OIII & SII
The narrowband data was mapped as follows - SII to red, HII to green & OIII to blue (SHO palette).

Closeup of Bubble:

""I am just learning to notice the different colors of the stars, and already begin to have a new enjoyment."
- Maria Mitchell, Astronomer

Object Types Included

Nebula (Dark)
X Nebula (Emission)
Nebula (Molecular Cloud)
Nebula (PN)
Nebula (Reflection)
Nebula (SNR)
Star Cluster (Open)
Star Cluster (Globular)
Galaxy

Wizard Nebula (NGC 7380)

Object:

Location Constellation of Cepheus, at a declination of +58 degrees

Distance 8000 light years

Size 70 light years

Description This object consists of a broad emission nebula, many dark dust clouds and lanes, and an open star cluster. The object spans about 100 light years and is one of the few DSOs that has a nickname I like.

NGC 7380 is the designation for the star cluster at the center of the image, which was the original object discovered in this area by Caroline Herschel in 1787. In astrophotography images like this one, this cluster is overshadowed by the magnificent surrounding emission complex.

I love how all of the towering dust cloud pillars point back to the cluster at the center, the result of the tremendous stellar winds that are shaping these dust clouds. Scientists believe that this kind of nebula is the birthplace of many stars, formed when the molecular clouds begin to collapse and fragment under their own gravity. I also like the details and shading of the bluish emission section which rises from the center of the cluster.

The "Wizard" nickname is perfect for this object. The wizard is sitting on his throne and looking right in this profile view. His head, at upper left, is highlighted by his pointy cap and nose. His eye and beard are clearly seen. I originally thought it could be a female wizard, but then I saw the beard. His hands and feet (complete with cool slippers) can also be discerned with some imagination. He must be thinking hard, because I can also see some deep and intense thoughts rising up and out from his pointy cap.

Image Acquisition:

Date Oct 11, 2023

Exp. Time 4 hours

Telescope Celestron RASA 11" V2
Camera ZWO ASI6200MM Pro
Mount Astro-Physics Mach1 GTO
Filters Broadband (for stars) - Red, Green & Blue
Narrowband (for nebula) - HII, OIII & SII
The narrowband data was mapped as follows - SII to red, HII to green & OIII to blue (SHO palette).

"Astronomy is useful because it raises us above ourselves; it is useful because it is grand; …. It shows us how small is man's body, how great his mind, since his intelligence can embrace the whole of this dazzling immensity, where his body is only an obscure point, and enjoy its silent harmony." - Henri Poincare, Physicist

Object Types Included

Nebula (Dark)
X   Nebula (Emission)
Nebula (Molecular Cloud)
Nebula (PN)
X   Nebula (Reflection)
Nebula (SNR)
Star Cluster (Open)
Star Cluster (Globular)
Galaxy

Trifid Nebula (M20)

Object:

Location   Constellation of Sagittarius, at a declination of -23 degrees

Distance   9000 light years

Size   70 light years

Description This object is a spectacular combination of an open star cluster, a reddish hydrogen emission nebula, a bluish reflection nebula and dark dust lanes.

Its nickname Trifid means 'divided into three lobes', although it looks more like five to me (a central lobe and 4 surrounding it). The brightest aspect of this object is near the center of the reddish nebula - a triple star system composed of three extremely hot stars. This triple star system is surrounded by a cluster of approximately 3000 stars, only a few of which can be seen in this image. The nebula lies in front of the central region of our galaxy, such that many stars comprising the Milky Way's galactic plane are seen in the image background.

My favorite aspects of this image are:

- The flowing wispy dark brown dust lanes.

- The pillars of star-forming dust, especially seen towards the bottom of the emission nebula.

- The "3-D" look of the bottom right surface of the reddish emission sphere, where it takes on a bit of an orange cast.

Image Acquisition:

Date   Aug 18-20, 2020

Exp. Time 7 hours

Telescope Celestron EdgeHD 11"
Camera   ZWO ASI6200MM Pro
Mount   Astro-Physics Mach1 GTO
Filters   Broadband - Lum, Red, Green & Blue
               Narrowband - HII - A hydrogen filter was used to bring out the HII gas

""We are all in the gutter, but some of us are looking at the stars."
- written by Oscar Wilde, sung by Chrissie Hynde

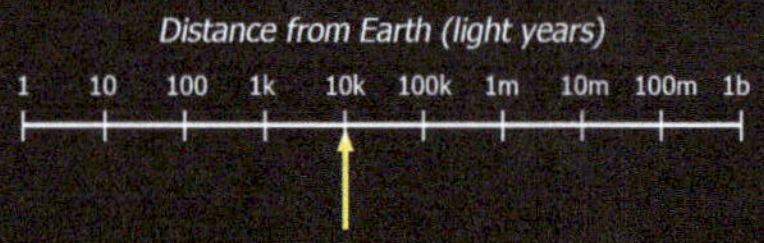

Distance from Earth (light years)

1   10   100   1k   10k   100k   1m   10m   100m   1b

Object Types Included

Nebula (Dark)
X Nebula (Emission)
   Nebula (Molecular Cloud)
   Nebula (PN)
   Nebula (Reflection)
   Nebula (SNR)
   Star Cluster (Open)
   Star Cluster (Globular)
   Galaxy

Pac-Man Nebula (NGC 281)

Object:

Location   Constellation of Cassiopeia, at a declination of +57 degrees

Distance   10,000 light years

Size   100 light years

Description This object with the 1980s nickname is a large emission nebula. The open star cluster at the center of the image, IC 1590, is ionizing and illuminating the surrounding gas. The large 8th magnitude star at the center, HD 5005AB, has four companion stars at distances between 1 and 16 arcsec.

Strong stellar winds from these stars are sculpting numerous elephant trunk projections along the outside of the nebula, with all of the trunks pointing towards the central star cluster. Numerous Bok globules are seen in silhouette against the nebula as well as a large dust cloud, also eroded by stellar winds. It is believed that star formation is occurring in these dense clouds of dust.

I especially like the small, intricate black cloud structures just left of the central cluster. I also find the large horizontal cloud structure below the cluster to be interesting. The erosion of this cloud appears to be symmetric from our perspective, which suggests to me that this cloud is between us and the cluster and is not immediately below it.

Image Acquisition:

Date   Dec 20, 2023

Exp. Time   4 hours

Telescope   Celestron RASA 11" V2
Camera   ZWO ASI6200MM Pro
Mount   Astro-Physics Mach1 GTO
Filters   Broadband (for stars) - Red, Green & Blue
                Narrowband (for nebula) - HII, OIII & SII
                The narrowband data was mapped as follows - SII to red, HII to green & OIII to blue (SHO palette).

"Do not look at stars as bright spots only. Try to take in the vastness of the universe." - Maria Mitchell, Astronomer

Distance from Earth (light years)

1   10   100   1k   10k   100k   1m   10m   100m   1b

Object Types Included

Nebula (Dark)
X Nebula (Emission)
Nebula (Molecular Cloud)
Nebula (PN)
Nebula (Reflection)
Nebula (SNR)
Star Cluster (Open)
Star Cluster (Globular)
Galaxy

Tadpoles Nebula (IC 410)

Object:

Location Constellation of Auriga, at a declination of +33 degrees

Distance 12,000 light years

Size 100 light years

Description This object, nicknamed the Tadpoles Nebula, is an emission nebula illuminated by the central open cluster of hot stars known as NGC 1893. These stars not only illuminate the nebula, but also emit strong stellar winds that have sculpted it.

At the center of the image, what were once two thick, majestic pillars of cold gas and dust have been so heavily eroded by the stellar winds that they now resemble tadpoles. A closeup of the tadpoles is shown below. The leading globules of the pillars still remain and represent the tadpoles heads, pointed at the star cluster. Radiation from the star cluster evaporates gas from the pillars, ionizes it, and blow it back against the globules, forming bright ridges of ionized gas. The tadpole pillars are about 6 arc-minutes long in our apparent view, which corresponds to a distance of about 15 light years.

The tadpoles are the highlight of this object, but I also like the numerous dark nebulae and billowing, colorful dust cloud shapes.

Image Acquisition:

Date Mar 4-5, 2023

Exp. Time 4 hours

Telescope Celestron RASA 11" V2
Camera ZWO ASI6200MM Pro
Mount Astro-Physics Mach1 GTO
Filters Broadband (for stars) - Red, Green & Blue
Narrowband (for nebula) - HII, OIII & SII
The narrowband data was mapped as follows - SII to red, HII to green & OIII to blue (SHO palette).

Closeup of Tadpoles:

"Space is for everybody. It's not just for a few people in science or math, or for a select group of astronauts. That's our new frontier out there, and it's everybody's business to know about space."
- Christa McAuliffe, Teacher and Challenger Astronaut

Object Types Included

  Nebula (Dark)
X Nebula (Emission)
  Nebula (Molecular Cloud)
  Nebula (PN)
  Nebula (Reflection)
  Nebula (SNR)
  Star Cluster (Open)
  Star Cluster (Globular)
  Galaxy

Lion Nebula (Sh2-132)

Object:

Location   Constellations of Cepheus and Lacerta, at a declination of +56 degrees

Distance   12,000 light years

Size   200 light years

Description   This object is a complex large emission nebula that has so many interesting features - dark nebulae, bright stars, billowing dust clouds, emission areas, and unusually bright streaks (both straight and curving).

The large orange star on the left edge of the image is one of the largest known stars in existence, with a diameter larger than the orbit of Jupiter. It is not certain whether this star, the hypergiant RW Cephei, is associated with the nebula or is just in the same field of view.

Two Wolf-Rayet stars, WR 152 and WR 153, are doing most of the ionization of the nebula, despite not being the brightest stars in the nebula. WR 152 is located in the center of the bluish area in the Lion's legs, while WR 153 is located in the center of the head of the Lion.

The most fascinating aspect of this image to me is the wide arcing fronts of two expanding gas bubbles. Surprisingly, the Wolf-Rayet stars described above are not the driving mechanism for the bubbles, as they are for other familiar objects such as Thor's Helmet, the Crescent Nebula, and Sh2-308.

One large bubble appears to be driven by the star cluster Berkeley 94, centered just below the bright orange star. The other large bubble encloses two dark nebulae towards the upper right, LDN 1150 and LDN 1154. It seems to me that the latter bubble could be driven by the apparent star cluster just to the left of these dark nebulae, but I could find no evidence in the literature that this grouping of stars is even a cluster.

Image Acquisition:

Date   Aug 20-21, 2023

Exp. Time   4 hours

Telescope   Celestron RASA 11" V2
Camera   ZWO ASI6200MM Pro
Mount   Astro-Physics Mach1 GTO
Filters   Broadband (for stars) - Red, Green & Blue
                  Narrowband (for nebula) - HII, OIII & SII
                  The narrowband data was mapped as follows - SII to red, HII to green & OIII to blue (SHO palette).

"Thus the heavens and the earth were completed in all their vast array." - Genesis 2:1

Distance from Earth (light years)

1   10   100   1k   10k   100k   1m   10m   100m   1b

Object Types Included
X   Nebula (Dark)
    Nebula (Emission)
    Nebula (Molecular Cloud)
    Nebula (PN)
    Nebula (Reflection)
    Nebula (SNR)
    Star Cluster (Open)
    Star Cluster (Globular)
    Galaxy

Mandrill Nebula (NGC 2467)

Object:

Location Constellation of Puppis, at a declination of -26 degrees

Distance 13,000 light years

Size 190 light years

Description This object is an emission nebula located in the southern constellation of Puppis at a declination of -26 degrees. The nebula consists of huge clouds of gas and dust sprinkled with bright stars and star clusters. It is unusual in that it consists of several superimposed nebula of different distances, which has led to large discrepancies in the reported distance of this object. The nebula is also known as Sh2-311 and Gum 9.

This area, which is believed to be just a few million years old, has been described as a very active stellar nursery.

Haffner 19 is the open star cluster surrounded by a distinct orange circular region, bordered in red hydrogen gas, located just above the center of the image. That orange region is a good example of a Strömgren sphere of ionized hydrogen gas, the most famous example of which is the Rosette Nebula (p. 52).

Haffner 18 is the larger cluster of about 50 stars at the center of the image. Just below this cluster, a young star can be seen still surrounded by its small circular birth cocoon of orange gas. This also looks like another Strömgren sphere to me, although I could not confirm this. A closeup of this region is shown below. Throughout the nebula are dark lanes of dust which are believed to be feeding the process of forming new stars.

Image Acquisition:

Date Feb 5, 2023

Exp. Time 4 hours

Telescope Celestron RASA 11" V2
Camera ZWO ASI6200MM Pro
Mount Astro-Physics Mach1 GTO
Filters Broadband (for stars) - Red, Green & Blue
Narrowband (for nebula) - HII, OIII & SII
The narrowband data was mapped as follows - SII to red, HII to green & OIII to blue (SHO palette).

Closeup of Central Region:

"I'm sure the universe is full of intelligent life. It's just been too intelligent to come here."
- Arthur C. Clarke, Science Fiction Writer

Object Types Included

   Nebula (Dark)
X Nebula (Emission)
   Nebula (Molecular Cloud)
   Nebula (PN)
   Nebula (Reflection)
   Nebula (SNR)
   Star Cluster (Open)
   Star Cluster (Globular)
   Galaxy

Flying Dragon Nebula (Sh2-114)

Object:

Location   Constellation of Cygnus, at a declination of +39 degrees

Distance   13,000 light years

Size   240 light years

Description   This object is a faint emission nebula located in the constellation of Cygnus at a declination of +39 degrees. Surprisingly little has been written about this beautiful object, so its nature remains a mystery. Its wispy filaments resemble a supernova remnant but no such remnant has been identified for this location.

I was surprised to find no papers or research on the origin of this object. Certainly this filamentary nebula has a story to tell - hopefully somebody will tell it soon.

The tiny bright red object above and slightly right of this nebula is the quadrupolar planetary nebula Lan 384, also known as Kn 26. A closeup of this PN is shown below.

Image Acquisition:

Date   Aug 17, 2023

Exp. Time   2 hours

Telescope   Celestron RASA 11" V2
Camera   ZWO ASI6200MM Pro
Mount   Astro-Physics Mach1 GTO
Filters   Broadband - Red, Green & Blue
Narrowband - HII - A hydrogen filter was used to bring out the HII gas

Closeup of Lan 384 Planetary Nebula:

"Then another sign appeared in heaven: an enormous red dragon with seven heads and ten horns and seven crowns on its heads. Its tail swept a third of the stars out of the sky and flung them to the earth." - Revelation 12:3-4

Object Types Included

Nebula (Dark)
X Nebula (Emission)
Nebula (Molecular Cloud)
Nebula (PN)
Nebula (Reflection)
Nebula (SNR)
Star Cluster (Open)
Star Cluster (Globular)
Galaxy

Thor's Helmet (NGC 2359)

Object:

Location Constellation of Canis Major, at a declination of -13 degrees

Distance 15,000 light years

Size 180 light years

Description This object, also known as Sharpless 2-298 and Gum 4, is an emission nebula located 12,000 light years away in the constellation of Canis Major at a declination of -13 degrees. The full extent of the nebula spans about 25 arc-minutes in our apparent view, about the size of our full moon.

This is one of my favorite deep sky objects. It has many nicknames, including Thor's Helmet, Flying Eye, and Duck Head Nebula. The nicknames should have stopped at Thor's Helmet.

The central star here is the Wolf-Rayet star WR7, the bright star seen near the center of the bubble. As we have seen many times before in this book, these are massive stars which have high temperatures, are highly luminous, and have a high rate of mass loss as they expel outer layers towards the end of their lives, resulting in beautiful nebula.

A closeup of the bubble is shown below. I used a longer focal length telescope with different filters for the image below, so the details are better and the colors a bit different than the full image. In both images, though, you will see something unusual - the filamentary sphere that is expanding outward in color from the central star is primarily white in color. This is because there are equal levels of hydrogen and oxygen in that region, instead of one gas being dominant as usually is the case.

Image Acquisition:

Date Dec 17, 2023

Exp. Time 4 hours

Telescope Celestron RASA 11" V2
Camera ZWO ASI6200MM Pro
Mount Astro-Physics Mach1 GTO
Filters Broadband (for stars) - Red, Green & Blue
Narrowband (for nebula) - HII, OIII & SII

Closeup of Central Region:

"Equipped with his five senses, man explores the universe around him and calls the adventure Science."
- Edwin Hubble, Astronomer

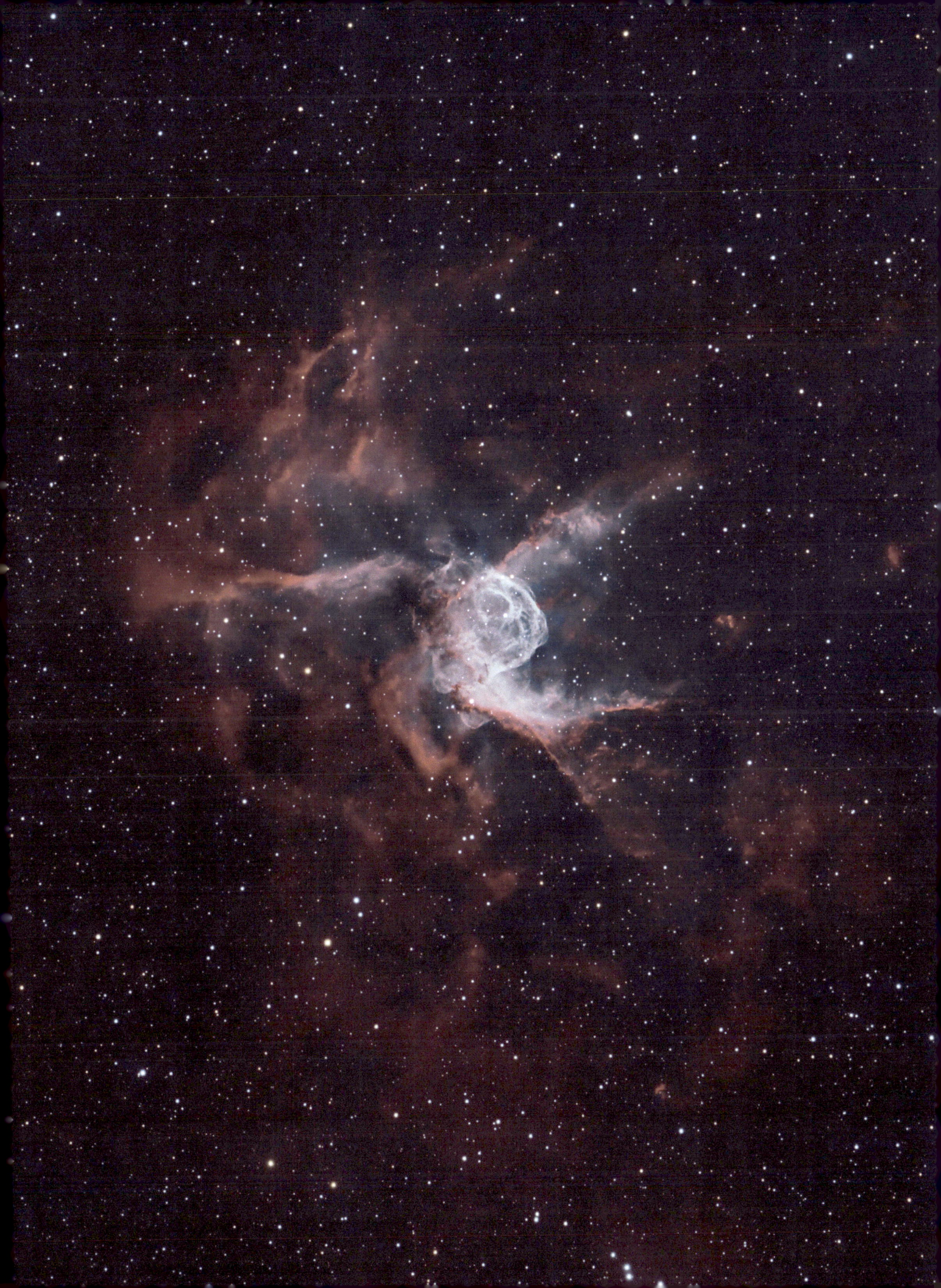

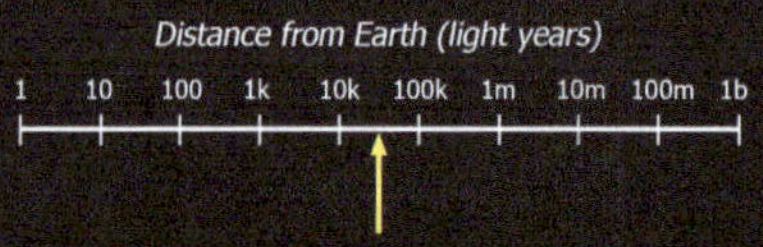

Object Types Included

Nebula (Dark)
Nebula (Emission)
Nebula (Molecular Cloud)
Nebula (PN)
Nebula (Reflection)
Nebula (SNR)
Star Cluster (Open)
X   Star Cluster (Globular)
Galaxy

Great Hercules Star Cluster (M13)

Object:

Location Constellation of Hercules, at a declination of +36 degrees

Distance 25,000 light years

Size 220 light years

Description This popular object is one of the sky's best globular clusters, located 25,000 light years away in the constellation of Hercules at a declination of +36 degrees.   It is bright enough to be seen as a fuzzy star with the naked eye, in dark skies. A closeup of the cluster is shown below.

there are two main types of star clusters in the sky. Globular clusters, such as this one, are tight groups of up to a millions of old star which are gravitationally bound. Open clusters, such as the Pleiades (p. 6), are loosely clustered groups of stars, containing fewer than a few hundred young stars.

This globular cluster contains about 300,000 stars and is about 220 light-years in diameter. It is considered to be the finest globular star cluster in the northern half of the heavens. In total, our Milky Way galaxy has about 160 of these globular clusters.

Note that the stars of this cluster look quite a bit smaller than some of the bright stars surrounding the cluster in the full image. That is because those brighter stars are much closer to us than this cluster.

Looking closely, you will see a small white galaxy in the top left quadrant of the full image. This galaxy is much further away than M13, about 4000x further at 100 million light years away.

Image Acquisition:

Date April 17, 2023

Exp. Time 1 hours

Telescope Celestron RASA 11" V2
Camera ZWO ASI6200MM Pro
Mount Astro-Physics Mach1 GTO
Filters Broadband - Lum, Red, Green & Blue

Closeup of M13 core:

"He determines the number of the stars and calls them each by name." - Psalm 147:4

Object Types Included

Nebula (Dark)
Nebula (Emission)
Nebula (Molecular Cloud)
Nebula (PN)
Nebula (Reflection)
Nebula (SNR)
Star Cluster (Open)
Star Cluster (Globular)
X Galaxy

Andromeda Galaxy (M31)

Object:

Location　Constellation of Andromeda, at a declination of +41 degrees

Distance　2,500,000 light years

Size　150,000 light years

Description　Now, for the first time in this book, we lift our gaze beyond our own Milky Way galaxy to reach out to distant worlds. Just 100 years ago, everyone believed that our galaxy was the only one in the universe. Now, not only do we know about millions (perhaps billions) of distant galaxies, we can see and image them from our own backyards.

This object is the beautiful and massive Andromeda Galaxy, located only 2.5 million light years from earth in the constellation of its own name. It is the closest spiral galaxy to us and the farthest object that we can see with our naked eye.

Its apparent size is about 6 times the width of our full moon. The disk contains many billions of suns. The massive black hole at the center of the galaxy is not visible. NGC206, the brightest star cloud in the galaxy, is visible at lower right. Two satellite galaxies are in the same view - M110 to the upper right and M32 to the lower left.

Over 500 globular clusters have been identified as belonging to M31.

Many dark dust lanes are evident from near the core all the way out to the outer ring. It is interesting to me how the dust lanes on the near (right) side of the galaxy disk show up much more prominently than the dust lanes on the far (left) side. The far side dust lanes are obscured by the misty cloud of stars in the galaxy halo.

My favorite aspect of this image is the many star clusters and Ha regions faintly visible in the outer region of the galaxy disk, especially top left and lower right.

Andromeda and our Milky Way galaxy are moving towards each other on a collision course. The good news is that the collision is billions of years away.

Image Acquisition:

Date　Dec 27, 2022

Exp. Time　2 hours

Telescope　Celestron RASA 11" V2
Camera　ZWO ASI6200MM Pro
Mount　Astro-Physics Mach1 GTO
Filters　Broadband - Lum, Red, Green & Blue

"In the year 3,000,002,012 the Andromeda Galaxy may collide with our Milky Way. At first this sounds miserable, like a collision of two bird flocks. But galaxy members fly farly, not tip to tip. In a galactic collision the stars do not actually collide—as with crisscrossing marching bands, only the interstices collide. (Oh to be like a galaxy, to mingle without wrecking. But then we would have to be composed of so much more sky.) The spaces between stars are so wide that thousands of galaxies have to converge before the stars will crash." — Amy Leach

Object Types Included

Nebula (Dark)
Nebula (Emission)
Nebula (Molecular Cloud)
Nebula (PN)
Nebula (Reflection)
Nebula (SNR)
Star Cluster (Open)
Star Cluster (Globular)
X Galaxy

Bodes & Cigar Galaxies (M81 & M82)

Object:

Location   Constellation of Ursa Major, at a declination of +69 degrees

Distance   12,000,000 light years

Size   92,000 light years (M81)

Description   This image captures two magnificent galaxies in the same field of view, located about 12 million light years away in the constellation of Ursa Major. M81, the lower galaxy, is a grand design spiral galaxy, while M82 is an irregular galaxy. Scientists believe that these two galaxies are interacting gravitationally. M81 and M82 lie in an region with extensive IFN (integrated flux nebulae), seen as faint white gaseous nebulae in the image.

At a diameter of 92,000 light years, M81 (the bottom galaxy) is slightly smaller in size than our Milky Way. This galaxy was first discovered by Johann Bode in 1774 and is sometimes referred to as "Bode's Galaxy", one of the few galaxies named after an individual.

Seen slightly just left of M81 is the dwarf irregular blue galaxy Holmberg IX, designated PGC 28757. This galaxy is a satellite galaxy of M81. A 2006 paper by Sabbi et al described this galaxy as the nearest young galaxy to earth, with an age of 200 million years.

M82 (the top galaxy) has a fascinating, irregular, distorted disk due gravitational influence from its large neighbor, M81. M82 is about half the size of M81. A closeup of M82 is shown below.

Our view of M82 is almost edge-on, at an inclination angle of 77 degrees. It is also known as Arp 337. The structure of this galaxy is so unusual that Dr. Halton Arp couldn't classify it, and it fell into the last category of the catalog called "Miscellaneous". It is called a starburst galaxy because it is undergoing a burst of new star formation. Through powerful winds from massive emerging stars, this burst of star formation in M82 is driving a huge outflow of reddish hydrogen gas from its galaxy core unlike that of any other galaxy. I like the dark lanes of brown dust running in strange patterns around the core of this unique galaxy.

Image Acquisition:

Date   Jan 13, 2024

Exp. Time   2 hours

Telescope   Celestron RASA 11" V2
Camera   ZWO ASI6200MM Pro
Mount   Astro-Physics Mach1 GTO
Filters   Broadband - Lum, Red, Green & Blue

Closeup of M82:

"The night is even more richly colored than the day. . . . If only one pays attention to it, one sees that certain stars are citron yellow, while others have a pink glow or a green, blue and forget-me-not brilliance. And without my expiating on this theme, it should be clear that putting little white dots on a blue-black surface is not enough."
- Vincent van Gogh, Painter

Object Types Included

Nebula (Dark)
Nebula (Emission)
Nebula (Molecular Cloud)
Nebula (PN)
Nebula (Reflection)
Nebula (SNR)
Star Cluster (Open)
Star Cluster (Globular)
X Galaxy

Whirlpool Galaxy (M51)

Object:

Location   Constellation of Canes Venatici, at a declination of +47 degrees

Distance   25,000,000 light years

Size   80,000 light years

Description   This Messier Object, also known as NGC 5194, is a face-on Seyfert grand design spiral galaxy located 25 million light years away in the constellation of Canes Venatici at a declination of +47 degrees. It is a magnitude 8.4 galaxy which spans 10 arc-minutes in our apparent view. This corresponds to a diameter of 80,000 light years, much smaller than our Milky Way. M51 is the 14th brightest galaxy in the sky.

The Whirlpool Galaxy is one of my favorite night sky objects. Viewed face-on from earth, the graceful, symmetric arms contain flowing dark lanes, blue star clusters and pink hydrogen star-forming regions. VV rows (straight arm segments) are seen in the grand design spiral arms.

The disks of grand design spiral galaxies are symmetric, such that the disk looks the same if it is rotated 180 degrees. Focus on just M51 (ignore the companion) and turn the book upside down. You will see that the disk looks exactly the same either way.

The companion galaxy seen left of M51 is NGC 5195. It may appear to you, as it did to scientists not that long ago, that these 2 galaxies are interacting. And they are. But the interaction is much less than it seems. The companion galaxy is actually behind M51. You can see that the companion galaxy is slightly disturbed, looking at its asymmetric outer halo of faint stars. However, M51 does not appear disturbed at all. The M51 dust lane at the lower left is silhouetted in front of the companion's core.

Numerous encounter theories have been proposed for those 2 objects. The sequence that I like the best is from the groundbreaking galaxy encounter simulation work by Toomre and Toomre. Their work indicates that the companion galaxy has passed M51 from top right to bottom left, just grazing the left edge of M51. The smaller galaxy has now passed behind M51, but the gravitation influences are still apparent in the distorted star halo of the companion. The two galaxies may eventually merge long in the future after several more "dances" around the sky, or their relative velocities may be different enough to continue taking them away from each other without merging.

Image Acquisition:

Date   April 6-7, 2022

Exp. Time   5 hours

Telescope   Celestron EdgeHD 11"
Camera   ZWO ASI294MM Pro
Mount   Astro-Physics Mach1 GTO
Filters   Broadband - Lum, Red, Green & Blue

"The spiral in a snail's shell is the same mathematically as the spiral in the Milky Way galaxy, and it's also the same mathematically as the spirals in our DNA. It's the same ratio that you'll find in very basic music that transcends cultures all over the world." - Joseph Gordon-Levitt

Object Types Included
Nebula (Dark)
Nebula (Emission)
Nebula (Molecular Cloud)
Nebula (PN)
Nebula (Reflection)
Nebula (SNR)
Star Cluster (Open)
Star Cluster (Globular)
X Galaxy

Bubble Galaxy (NGC 3521)

Object:

Location   Constellation of Leo, at a declination of +0 degrees

Distance   30,000,000 light years

Size   100,000 light years

Description   This object is a spiral galaxy located 30 million light-years away in the constellation of Leo at a declination of 0 degrees. It is about 100,000 light years in diameter and spans 11 arc-minutes in our apparent view. The galaxy is 30 degrees from edge-on, which allows us to see characteristics which are typical for both edge-on and face-on galaxies. This is one of my favorite unsung galaxies.

The galaxy has a number of interesting qualities - the flocculent nature of the multiple arms, the blue star clouds and pink Ha regions, the detailed dust lanes (including one at the bottom which appears to veer out of plane), and the bright central core.

But there are two things which make this galaxy even more special. The first is the galaxy's roughly spherical halo of stars which encompasses the disk. Once your eyes adjust to it a bit, the "bubble" gives the galaxy a 3-D type of appearance. It looks to me like the whole galaxy is steaming. Scientists believe that this tidal stream of stars is the result from one or more galaxies which have merged with NGC 3521 long ago.

The second thing is difficult to see at first, but then obvious once you are aware of it. The bright disk of the inner region is hexagonal in shape. The outer disk also appears a bit hexagonal as well. Hexagonal shapes occur many places throughout nature such as in bee honeycombs and in the cloud pattern around the north pole of the planet Saturn, but those seen in galaxies are the largest scale hexagons in the universe. The hexagonal shape of the inner region is outlined in yellow below.

Image Acquisition:

Date   April 18, 2023

Exp. Time   3.5 hours

Telescope   Celestron EdgeHD 11"
Camera   ZWO ASI294MM Pro
Mount   Astro-Physics Mach1 GTO
Filters   Broadband - Lum, Red, Green & Blue

Hexagonal Shape of Bright Inner Region:

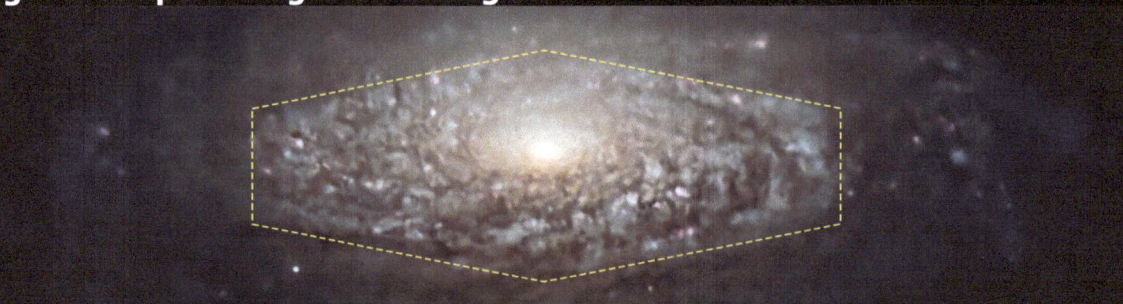

"The lunar flights give you a correct perception of our existence. You look back at Earth from the moon, and you can put your thumb up to the window and hide the Earth behind your thumb. Everything you've ever known is behind your thumb, and that blue-and-white ball is orbiting a rather normal star, tucked away on the outer edge of a galaxy."
- Jim Lovell

1   10   100   1k   10k   100k   1m   10m   100m   1b

Object Types Included
Nebula (Dark)
Nebula (Emission)
Nebula (Molecular Cloud)
Nebula (PN)
Nebula (Reflection)
Nebula (SNR)
Star Cluster (Open)
Star Cluster (Globular)

Leo Galaxy Triplet (M65/M66/NGC 3628)

Object:

Location	Constellation of Leo, at a declination of +13 degrees
Distance	41,000,000 light years (M65)
Size	110,0000 light years (M65)

Description This object is the famous Leo Triplet of galaxies, located in the constellation of Leo. I had previously thought that all 3 of these magnitude 9 galaxies are at the same distance from us, but that is incorrect. NGC 3628 (at top) and M66 (at lower left) are about the same distance, at 35 million light years, which explains the gravitation interaction that we see in their disks and star streams. M65 (lower right) is slightly further away at 41 million light years.

NGC 3628, nicknamed the Hamburger Galaxy, is the edge-on spiral galaxy at top. It spans 17 arc-minutes in our apparent view, which corresponds to a diameter of 160,000 light years. It is rare that a non-Messier object outshines 2 Messier objects in the same frame, but that is the case here because of its beautiful dust lane and 400,000 light year long star stream tail. The tail is shown in the closeup below.

The broad equatorial dust band of NGC 3628 obscures the galaxy's bright central region and hides most of the bright young stars in its spiral arms. Gravitational interactions with M65 are believed to be responsible for the extended flare and warp of this spiral's disk, as well as the starstream tail. The gravitational pull has most likely tipped the plane of NGC 3628 and made its central dust lane "wobbled" in appearance. A small satellite dwarf galaxy is seen just below NGC 3628.

M66 is the face-on galaxy at lower left. It has a diameter of 90,000 light years and is oriented about 25 degrees from edge-on in our apparent view. While all of these galaxies exhibit prominent dust lanes sweeping along their broad spiral arms, M66 is particular interesting in the contrast between the pinkish hues of hydrogen gas in star forming regions and the bluish hues of the young blue star clusters. It also displays a slightly odd non-symmetric structure, likely due to gravitation interaction with NGC 3628.

M65 is the undisturbed galaxy at lower right. It has a Milky Way like diameter of 110,000 light years and is oriented about 15 degrees from edge-on in our apparent view. I find this galaxy fascinating because of the symmetric brightness undulations as you move from the top of the galaxy to the bottom. The dark lanes appear as six dark circles as you move across the disk, interrupted by the glowing core.

Image Acquisition:

Date	Jan 26, 2023
Exp. Time	2 hours
Telescope	Celestron RASA 11" V2
Camera	ZWO ASI6200MM Pro
Mount	Astro-Physics Mach1 GTO
Filters	Broadband - Lum, Red, Green & Blue

Closeup of NGC 3628 Star Stream Tail:

"In less than a hundred years, we have found a new way to think of ourselves. From sitting at the center of the universe, we now find ourselves orbiting an average-sized sun, which is just one of millions of stars in our own Milky Way galaxy."
- Stephen Hawking

Object Types Included

Nebula (Dark)
Nebula (Emission)
Nebula (Molecular Cloud)
Nebula (PN)
Nebula (Reflection)
Nebula (SNR)
Star Cluster (Open)
Star Cluster (Globular)
X   Galaxy

Antennae Galaxies (Arp 244)

Object:

Location   Constellation of Corvus, at a declination of -19 degrees

Distance   60,000,000 light years

Size   50,000 light years

Description   This Arp object, also known as NGC 4038 & 4039 as well as Caldwell 60 & 61, is a pair of colliding galaxies located 60 million light years away in the constellation of Corvus at a declination of -19 degrees.

The Arp Atlas of Peculiar Galaxies is a catalog of 338 peculiar galaxies produced by Dr. Halton Arp in 1966. Although most are tiny and hard to image clearly, they are some of the most interesting looking galaxies in the universe. This one, Arp 244, is one of the best.

I find this object to be one of the most visually fascinating of all of the deep sky objects and a great example of 2 merging galaxies. The galaxies are well along in the merger process and have each been so severely deformed by their gravitational interaction that it is hard to recognize the original shape of each galaxy.

This object is unusual due to the high intensity and extent of the pink regions. These regions are associated with active star formation, which has been initiated as a result of the gravitational interactions of this pair. The clouds of ionized hydrogen are dense and radiate a pink glow as they are lit up by the young stars nearby.

The two long tidal tails have a combined length of almost 400,000 light years, almost 3 times longer than our Milky Way galaxy. The long tails in this image are a great example of how tidal interactions from approaching galaxies tend to cause the tails to flow out away from each other.

Image Acquisition:

Date   May 25-27, 2022

Exp. Time   10.5 hours

Telescope   Celestron EdgeHD 11"
Camera   ZWO ASI294MM Pro
Mount   Astro-Physics Mach1 GTO
Filters   Broadband - Lum, Red, Green & Blue
Narrowband - HII   The narrowband HII data was used to bring out the star formation areas

"The heavens praise your wonders, O Lord." - Psalms 89:5

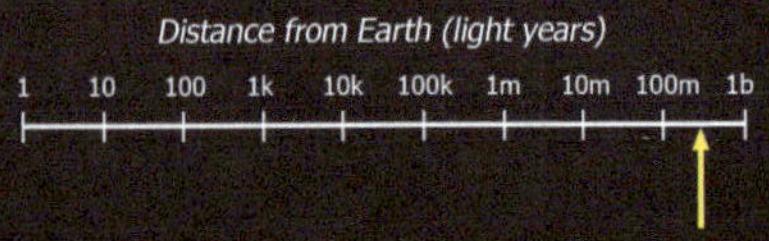

Object Types Included

Nebula (Dark)
Nebula (Emission)
Nebula (Molecular Cloud)
Nebula (PN)
Nebula (Reflection)
Nebula (SNR)
Star Cluster (Open)
Star Cluster (Globular)
X Galaxy

Coma Galaxy Cluster (Abell 1656)

Object:

Location   Constellation of Coma Berenices, at a declination of +28 degrees

Distance   300,000,000 light years

Size   6,000,000 light years (width of cluster)

Description   Abell galaxy clusters, named after Dr. George Abell who cataloged them, encompass massive space and distance and are the largest known gravitationally bound structures in the universe. Galaxy cluster images help us to appreciate the vastness of the huge universe which lies beyond the boundaries of our solar system.

Abell 1656 is a large galaxy cluster, nicknamed the Coma Cluster, located 300 million light years away in the constellation of Coma Berenices at a declination of +28 degrees. This cluster contains over 1000 galaxies. A closeup of the central region of this cluster is shown below, turned 90 degrees to better fit the image.

Since this cluster is 300 million light years away, that means that it takes 300 million light years for the light we are seeing from this cluster to travel to earth. This image that we are looking at, taken in 2024, captures what life was like in this galaxy cluster 300 million years ago. In many ways that is hard to believe, but that is what science and math tells us. The scope of the universe in both distance and time is hard to comprehend relative to our normal life on earth.

The 2 large elliptical galaxies which anchor this cluster are massive, each spanning a diameter of over twice that of our Milky Way. Elliptical galaxies look like big fuzzy spherical balls, with no arms or disk structure.

Plenty of spiral galaxies are seen here. The most visible (brightest) ones are edge-on to our view, since most of their starlight is concentrated in a such a small profile.

Image Acquisition:

Date   April 30, 2024

Exp. Time   3.5 hours

Telescope   Celestron EdgeHD 11"
Camera   ZWO ASI6200MM Pro
Mount   Astro-Physics Mach1 GTO
Filters   Broadband - Lum, Red, Green & Blue

Closeup of Central Region:

"Our galaxy, the Milky Way, is one of 50 or 100 billion other galaxies in the universe. And with every step, every window that modern astrophysics has opened to our mind, the person who wants to feel like they're the center of everything ends up shrinking."
- Neil deGrasse Tyson

Resources

The following resources compliment this book material:

Imm Astrobin Web Site - My Astrobin site has over 2500 DSO images which may be freely downloaded in full resolution, including all of the objects in this book. These images include many interesting objects which could not be included in this book, such as the Witch Head Nebula, the Diamond Ring Nebula, the Rose Galaxy, the Dolphin Nebula, and God's Hand. Along with each image is a written description of the object, complete technical data on the telescope/equipment setup, acquisition details such as exposure time and filter, and image comments from the Astrobin community. My Astrobin site can be viewed at https://www.astrobin.com/users/GaryI

Grayscale Astro Coloring Book - Do you love space? Relax, enjoy and learn as you color 18 images of true deep space objects. You will learn about each object first, then view an actual object image, and finally color a grayscale version of the object. You may choose to match the original object colors or be creative and come up with a novel new color palette. The book, Beautiful Space Objects - Volume 1, can be ordered from Amazon for $14.99.

If you enjoy space beauty, please check out my other astrophotography books on Amazon.

Scholarship

If you would like to help the next generation of astronomers, please consider a small donation to the Imm Astronomy Scholarship Fund, a tax-deductible donation account established to provide financial assistance to college students with an interest in astronomy.   100% of your donation goes to student scholarships. Scholarships are awarded annually in June.

Search for "imm astronomy scholarship fund" to find both the donor site and, if you are a student, the application site.

Conclusion

Congratulations, you are one of the few who have taken the time to view, understand and appreciate this incredible universe and its beauty! I hope that you have enjoyed seeing the objects in this book as much as I have enjoyed imaging and researching them.

Space is a magnificent place where children naturally have an interest and curiousity. Some of my favorite moments have involved teaching children about space and showing them the stars, moon and planets. It is easy for us to lose that sense of awe and wonder as we get older. Hopefully this book will help rekindle or reinforce an interest in space for you.

Please contact me at immgr@swbell.net if you have any questions, corrections, or suggestions for improvement, and particularly if you would like to see the completion of Volume 2 of this series, which is currently under development.

All images by Gary Imm
from Onalaska, TX, USA
(SQM Magnitude 20.8 - Bortle 4.5)